Indentation Techniques in Ceramic Materials Characterization

For information on ordering titles published by The American Ceramic Society, or to request a publications catalog, please contact our Customer Service Department at:

Customer Service Department
PO Box 6136
Westerville, OH 43086-6136, USA
614-794-5890 (phone)
614-794-5892 (fax)
info@ceramics.org

Visit our on-line book catalog at www.ceramics.org.

Ceramic Transactions
Volume 156

Indentation Techniques in Ceramic Materials Characterization

Proceedings of the International Symposium on Indentation Techniques in Ceramic Materials Characterization held at the 105th Annual Meeting of The American Ceramic Society, April 27-30, 2003, in Nashville, Tennessee

Edited by

Ahmad G. Solomah
SAC International Ceramics

Published by
The American Ceramic Society
PO Box 6136
Westerville, Ohio 43086-6136
www.ceramics.org

Proceedings of the International Symposium on Indentation Techniques in Ceramic Materials Characterization held at the 105th Annual Meeting of The American Ceramic Society, April 27-30, 2003, in Nashville, Tennessee

COVER PHOTO: "FIB images of the 100g load Vickers indent on the surface normal to the hot-pressing direction of the RL microstructure" is courtesy of R. J. Moon, Z-H Xie, M. Hoffman, P. R. Munroe, and Y-B Cheng appears as figure 7 in their paper "Application of Focused Ion Beam Miller in Indentation Fracture Characterization" which begins on page 49.

For information on ordering titles published by The American Ceramic Society, or to request a publications catalog, please call 614-794-5890.

4 3 2 1–07 06 05 04

ISSN 1042-1122
ISBN 1-57498-212-5

Contents

Preface

Indentation techniques have become widely used in the characterization of brittle solids due to their simplicity, cost effectiveness, rapidness, and maybe most importantly, the indenter itself can be used as a "mechanical" microprobe in thin films, interfaces and grain boundaries, and nanocomposites.

The aim of this International Symposium on Indentation Techniques in Ceramic Materials Characterization, which was held during the 105th Annual Meeting of The American Ceramic Society, April 27-30, 2003, in Nashville, Tennessee, was to discuss measurement techniques, reliability of, and problems associated with this fascinating testing method.

The application of the nanoindentation technique, as a new frontier in brittle solids characterization (i.e., thin film and nanocomposite materials), has been included in the scope of this symposium. Uses of indentation mechanics to analyze industrial and medical applications, (i.e., grinding, scribing and fatigue in dental ceramics), is included in this volume.

Speakers from Germany, Italy, Japan, Korea, USA, France, and Canada presented very interesting research results that have enlightened our knowledge and enriched our information about the usefulness of indentation techniques and their invaluable applications in a wide range of materials characterization. Accuracy of the "empirical" formulae used to calculate the fracture toughness of transformation-toughened zirconia ceramics was presented and discussed and the applicability of such empirical equations has been criticized due to their inconsistency and inappropriateness to be used for such phase-transformation materials. Application of Raman spectroscopy to determine the monoclinic fraction of transformed tetragonal zirconia grains around scratches and indents has been presented and included in this volume.

Finally, the open forum which was designed to exchange ideas and discuss the technical and analytical problems associated with indentation techniques has shed the light on many forgotten issues that were ignored over the last three decades of research.

Indeed, this symposium was a useful and informative for materials scientists who are involved in the research, development and applications of advanced ceramic materials and composites.

Finally, I would like to dedicate this volume to Hayne Palmour III, Tom Hare, Eric Merz, Reinhard Odoj, and Hansjoachim Matzke.

Ahmad G. Solomah

Indentation Techniques in Ceramic Materials Characterization

ACCURACY OF EMPIRICAL FORMULAE TO MEASURE THE TOUGHNESS OF TRANSFORMATION-TOUGHENED CERAMICS – HOW ACCURATE DO YOU NEED YOUR K_{IC}?

Ahmad G Solomah
SAC International Ceramics, Mississauga, Ontario Canada

ABSTRACT

Vickers micro-indentation technique has been applied to determine the fracture toughness or K_{IC} of two zirconia transformation-toughened ceramic materials: yttria-tetragonal zirconia polycrystals (Y-TZP, 2mole% Y_2O_3) and yttria-partially-stabilized zirconia (Y-PSZ, 5 mole% Y_2O_3). Both materials displayed two different indentation fracture behaviors, which were related to the stabilizer's contents, and to the grain size to some extent. The empirical formulae, that are available in the literature, were used to calculate the toughness values of such engineering ceramic materials that are expected to have high toughness values. Significant variations in K_{IC} values for the same material was found due to the problems associated with the original formulations of such empirical equations. It is concluded that new formulations must be developed to accommodate for the phase transformation process that changes the microstructure during crack propagation (stress induced tetragonal-to-monoclinic (t->m) phase transformation accompanied with an increase in lattice volume) and, subsequently, crack propagation shape and length change drastically. The stress induced t->m phase transformation in zirconia-toughened ceramics is a complex phenomenon, which is a chemi-thermodynamically-controlled process, and subsequently, it deserves more investigation to accurately be able to calculate reliable toughness values for their structural applications, especially under temperature and stress loading conditions.

INTRODUCTION

Transformation-Toughened Ceramics (TTC's) represent a new class of engineering materials that have shown their promises to be used in many high-technological applications as well as to replace their metallic counterparts in many industrial fields. The tetragonal-to-monoclinic (t->m) phase transformation is accompanied with a volume increase that creates significant compressive stresses at the crack tip and around the crack path, which will impede the crack propagation and, enhance the crack resistance of the materials, i.e., increase in their toughness (1).

Application of Vickers micro-indentation technique to study the indentation fracture and, subsequently, to calculate the fracture toughness of brittle solids, particularly glasses and ceramics, has become a routine testing technique since its conception in mid 1970's (2-21). The test has several advantages: (1) it can be used on small samples, (2) preparation of test specimens is relatively easy and simple, (3) it is a fast and cost effective test, (4) the Vickers diamond indenter used to produce the hardness indentation is a standard equipment, and, (5) it can be regarded as a "non"-destructive testing technique which can be used as a quality control method. The Vickers micro-indentation technique has been successfully applied to determine the fracture toughness, K_{IC}, of several brittle solid materials that include soda lime glasses (K_{IC} < 1 $MPa.m^{1/2}$), alumina (K_{IC} ~ 3-4 $MPa.m^{1/2}$) and hard metals like cemented carbides, e.g., WC-Co (K_{IC} ~ 12-16 $MPa.m^{1/2}$). Many empirical formulae were developed to calculate K_{IC} from indentation load, P (or hardness H), the indentation crack length, c, and the half-length of the indentation diagonal, a, for different crack configurations, i.e., median (half-penny), lateral or Palmqvist cracks. All the empirical formulae available in the literature were derived from the "spherical pressurized cavity" model, and assuming that the material is "well-behaved". The term "well-behaved" means that the material does not go through any phase change under stress application during the indentation process. It is well known that the tetragonal zirconia crystals (or phase) go through phase transformation to monoclinic symmetry as the crystals experience a stress. As a result of this stress and tetragonal-to-monoclinic phase transformation, an increase in the volume of the crystals (~ 3-5 vol.%) is developed and creates significant compressive stresses, as high as ~2-3 GPa. Such a high compressive stress zone, in front and around the crack path, will slow down or stop, thus increasing the crack resistance of the matrix (material) and subsequently, fracture toughness or KIc will increase dramatically. This process is called "Transformation Toughening." Despite this obvious and well-known phase change in tetragonal zirconia-based ceramics, which are considered "not well-behaved" materials, these empirical formulae were widely used to determine KIc without any conservation to their validity or consistency by many research groups all over the world. It is the purpose of this paper to present and to discuss the problems associated with the empirical formulae in calculating the "fracture" toughness of transformation-toughened zirconia-based ceramics using Vickers micro-indentation technique.

EXPERIMENTAL PROCEDURES

Pellets of yttria-tetragonal zirconia polycrystals (Y-TZP - 2 mol% Y2O3) and yttria-partially-stabilized zirconia (Y-PSZ – 5mole% Y_2O_3) were prepared by pressureless sintering technique. Their starting powders were prepared in our laboratory using a proprietary process that produced an ultra-fine and highly reactive sinterable powders. The pellets dimensions were 18 mm in diameter and 15 mm thickness. They were one surface polished to a mirror finish using 0.3 um diamond paste. The weight fraction of monoclinic, tetragonal and cubic phases of zirconia were determined using x-ray diffraction of the polished surfaces and applying the formulae given in the literature, in which the integrated intensities of the reflections (111)m, (111)t, (004)t, (400)t and (400)c are used. Polishing was terminated when the mole fraction of monoclinic phase was below 0.1% in Y-TZP materials after two consecutive polishing steps. In Y-PSZ ceramics, polishing was terminated when the mole fraction of cubic phase did not change after two consecutive polishing steps. No heat treatments or annealing were carried out after polishing was terminated.

Vickers micro-indentations were performed on the polished surfaces using a standard hardness testing equipment at ambient atmosphere (20 C and 50% relative humidity). The micro-indentation loads used were 49.05, 98.10, 196.20 and 294.30 N. The dwell time was 15-17 seconds and the elapsed time between unloading and crack length measurement was ~ 50 seconds. At least six micro-indentations were performed for each load selected. The average values of the half-diagonal of the indentation impression, a, and of the crack length, c, were used to calculate the hardness, H(i) and the toughness, KIc(i), for each micro-indentation of a specific load, Pi, respectively. The overall average values of hardness, H, and toughness, KIc, were calculated for each micro-indentation load from the six values of H(i)'s and KIc(i)'s, respectively.

The empirical formulae used to calculate the fracture toughness, KIc(i) of the Y-TZP and Y-PSZ transformation-toughened zirconia ceramics are compiled and listed in Table I (6-17), as they were developed over a time period of about two decades.

TABLE I Empirical formulae used to calculate fracture thoughness, K_{Ic}, of ceramic materials using Vickers indentation technique*

	Formulae	Author(s)	Ref.No.
(1)	$K_{Ic} = 0.1704\, Ha^{1/2} \log(4.5\, ^{a}/c)$	Evans & Wilshaw	6
(2)	$K_{Ic} = 0.057\, Ha^{1/2} (E/H)^{2/5} (^{c}/a)^{-3/2}$	Evans, et al	7,10
(3)	$K_{Ic} = 0.0303\, Ha^{1/2} (E/H)^{2/5} \log (8.4\, ^{a}/c)$	Blendel	8
(4)	$K_{Ic} = 0.0139\, (E/H)^{1/2}\, P\, c^{-3/2}$	Lawn, et al	9
(5)	$K_{Ic} = 0.016\, (E/H)^{1/2}\, P\, c^{-3/2}$	Anstis, et al	11
(6.a)	$K_{Ic} = 0.0711\, Ha^{1/2} (E/H)^{2/5} (^{c}/a)^{-3/2}$ for $^{c}/a > 2.5$	Niihara, et al	12
(6.b)	$K_{Ic} = 0,0193\, Ha^{1/2} (E/H)^{2/5} [(c-a)/a]^{-1/2}$ for $1.25 < {}^{c}/a < 3.5$	Niihara,et al	12
(7)	$K_{Ic} = 0,0782\, Ha^{1/2} (E/H)^{2/5} (^{c}/a)^{-1.56}$	Lankford	13
(8)	$K_{Ic} = 0.0098\, (E/H)^{2/3}\, P\, c^{-3/2}$	Laugier	14
(9)	$K_{Ic} = 0.035\, (E/H)^{1/4}\, P\, c^{-3/2}$	Tanaka	15
(10)	$K_{Ic} = 0.015\, (E/H)^{2/3} [(c-a)/a]^{-1/2}\, P\, c^{-3/2}$	Laugier	16
(11)	$K_{Ic} = 0.551\, \acute{\alpha}\, Ha^{1/2} (E/H)^{2/5} (c/a)^{(c/18a) - 1.51}$ where $\acute{\alpha} = 1/\alpha = 1/[14(1-8((4\nu-0.5)/(1+\nu))^{4})]$	Liang, et al	17

E = Elastic modulus
H = Hardness
P = Applied load
a = Half-indentation diagonal
c = Crack length measured from centre of indentation diagonal
ν = Poisson's ratio

* The value of the constant ϕ given in the original forms of these empirical equations, has been assumed to be 2.7.

RESULTS, ANALYSES AND DISCUSSIONS

A. Yttria-Tetragonal Zirconia Polycrystals (Y-TZP) Monolithic Materials

The x-ray diffraction (XRD) analyses carried out on the polished surfaces of all Y-TZP specimens confirmed the major phase is tetragonal zirconia symmetry and the monoclinic phase that resulted due to grinding and polishing processes, is less than 0.1%. Sequential polishing has proved that its necessity to remove most of the monoclinic phase and eliminate any near surface residual compressive stresses, due to t->m transformation, which would underestimate the crack length and subsequently, unrealistic higher toughness values would result (18-21). The average grain size of Y-TZP materials was about 0.3 um.

Examining the equations listed in Table I, one can notice that the fracture toughness, KIc, is proportional to the indentation load, P, and the crack length, c, of power of minus three-halves, i.e., KIc $Pc^{-3/2}$. Since KIc is a constant value of a specific material toughness, therefore, $Pc^{-3/2}$ is constant, i.e., P $c^{3/2}$ or c $P^{2/3}$. These constants given in the empirical formulae shown in Table I include the elastic modulus, E, the hardness, H, and numerical constants which were obtained from the least-square fits of the plots of log [KIc/(Ha ½)] [(H/E)2/5] versus log [c/a] for several ceramic materials whose KIc values were determined using conventional testing methods, e.g., single-edge notched bean (SENB) and double-cantilever beam (DCB). The significant differences between these empirical formulae are the numerical constants and the exponents of the ratio of [E/H]. Therefore, we have plotted the natural log of the crack length, c, versus the natural log of the micro-indentation load, P, to examine the slope of such a straight line fit, to validate the original assumption mentioned above. Regardless of what type of crack formed after indentation, Figure 1 shows that the slope of the least-square fit of the linear relationship between ln c and ln P is 0.589, which is significantly less than the 2/3 value indicated in the formulation of these empirical equations, i.e., 0.667. This indicates that the relationship between KIc, P and c, i.e., KIc $Pc^{-3/2}$, which is the base model of the empirical equations given in Table I, is not valid for Y-TZP monolithic tetragonal zirconia ceramic materials.

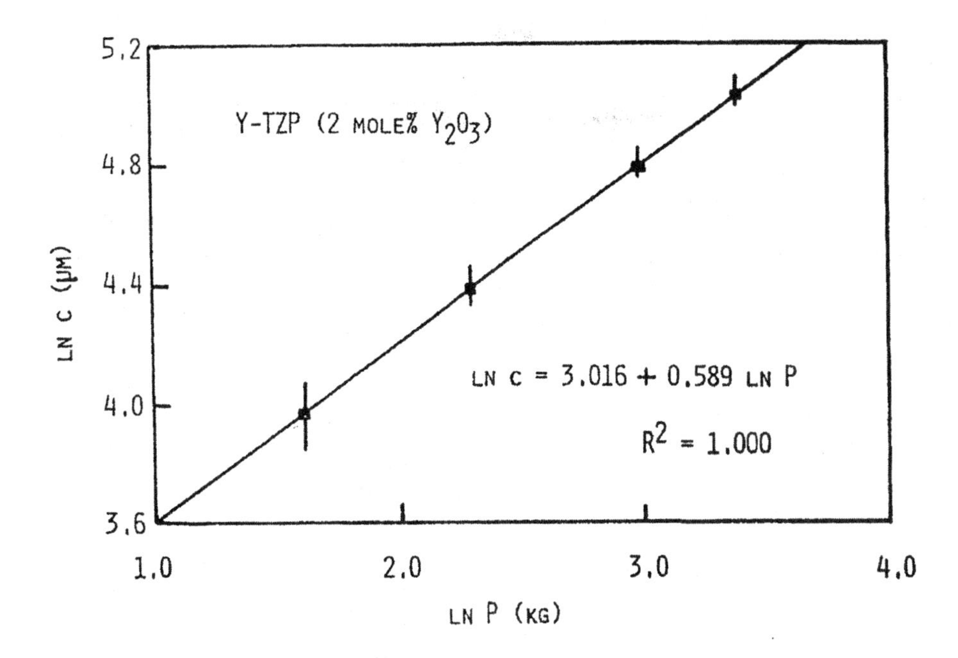

Figure 1. The relationship between the crack length, c, and indentation load, P, for yttria-tetragonal zirconia polycrystals (Y-TZP) ceramic materials.

Table II shows the fracture toughness values of Y-TZP monolithic ceramics as calculated using the equations given in Table I. The hardness, H, fracture toughness, KIc, (c/a) and Pc^-3/2 are shown as a function of micro-indentation load, P. It is evident that the harness value of Y-TZP monolithic ceramic materials is load-independent, which ranges between 12.1 and 12.3 GPa, a typical value of such a material. On the other hand, the toughness values vary drastically from one formula to another for a specific load, and it also changes with the indentation load, P, i.e., a load-dependent property, which is not valid physically. Also, the quantity Pc^-3/2 must be load independent, however, Table II shows that it is a load-dependent quantity, which is meaningless. These findings are quite interesting to prove that the empirical equations given in Table I are NOT valid to be used to calculate the "fracture" toughness of Y-TZP monolithic ceramic materials. It can also be said that the microstructure of such materials plays an important role in the phase transformation process that is highly dependent on the stabilizer distribution throughout the crystals (grains), their size and their orientation with respect to the crack propagation path. Also, the phase transformation that causes the chemithermodynamic changes in the t->m crystals, will affect the entropy and energy state of each phase. Therefore, the Y-TZP materials behave in a very different manner from the so-called "well behaved" or the "reference" materials, which were used in developing these empirical formulae.

TABLE II Hardness (H) and fracture toughness (K_{Ic}) of Yttria-tetragonal zirconia polycrystals (Y-TZP) as a function of indentation load (P)

P (N)	(c/a)	H (GPa)	$Pc^{-3/2}$ ($MN.m^{-3/2}$)	K_{Ic} ($MPa.m^{1/2}$)	Formula used
49.1	1.225	12.341	128.576	7.77	(1)
				10.48	(2)
				6.32	(3)
				7.33	(4)
				8.43	(5)
				10.09	(6.b)
				14.21	(7)
				8.27	(8)
				9.11	(9)
				26.64	(10)
				15.93	(11)
98.1	1.304	12.192	138.095	8.72	(1)
				11.34	(2)
				7.24	(3)
				7.91	(4)
				9.11	(5)
				10.29	(6.b)
				15.27	(7)
				8.95	(8)
				9.81	(9)
				24.85	(10)
				17.27	(11)
196.2	1.376	12.230	151.751	9.95	(1)
				12.49	(2)
				8.38	(3)
				8.68	(4)
				9.99	(5)
				11.01	(6.b)
				16.72	(7)
				9.81	(8)
				10.78	(9)
				24.47	(10)
				19.05	(11)
294.3	1.427	12.071	157.527	10.57	(1)
				12.96	(2)
				9.04	(3)
				9.07	(4)
				10.44	(5)
				11.39	(6.b)
				17.41	(7)
				10.27	(8)
				11.22	(9)
				24.21	(10)
				19.95	(11)

B. Yttria-Partially-Stabilized Zirconia (Y-PSZ) Ceramic Materials

The x-ray diffraction (XRD) analyses of the polished surfaces of Y-PSZ (5 mole% Y_2O_3) have confirmed the co-existence of both tetragonal and cubic phases of ZrO_2. No monoclinic phase was detected using routine x-ray diffraction analyses, i.e., < 0.1%). The average grain size of Y-PSZ materials was about 1 um.

The relationship between the crack length, c, and the micro-indentation load, P, was examined and it was found that the slope of the linear relationship between lnc and lnP is 0.678, which is in an excellent agreement with the aforementioned model, i.e., c P^2/3. Figure 2 shows the plot of such a relationship with the best-fit linear fit of regression coefficient of 0.998. This indicates that the empirical equations given in Table I can be used to calculate the "toughness" of Y-PSZ within the accuracy limit that has been mentioned by Evans and Charles, i.e., ~ < 30%. Table III shows the fracture toughness, KIc, values calculated using the empirical equations given in Table I, in addition to the hardness, H, and the quantity Pc^-3/2, as a function of indentation load, P. It is also shown that the quantity Pc^-3/2 is fairly independent of indentation load, i.e., within measurement limit 5~7%. It is also interesting to find out that for a specific equation, the value of the fracture toughness, KIc, is fairly consistent over the wide range of micro-indentation load, P, applied, i.e., + - 5%.

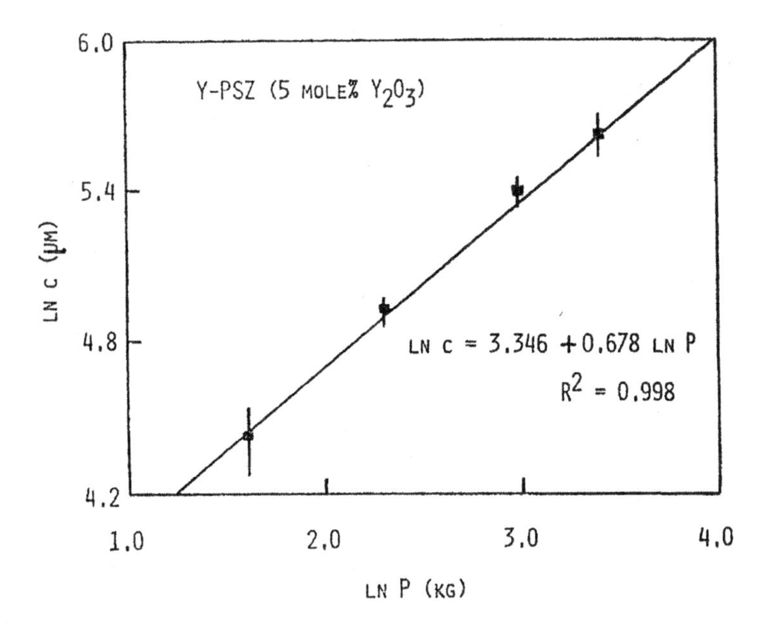

Figure 2. The relationship between the crack length, c, and indentation load, P, for yttria-partially stabilized zirconia (Y-PSZ) ceramic materioals.

TABLE III Hardness (H) and fracture toughness (K_{Ic}) of Yttria-partially-stabilized zirconia (Y-PSZ (5 mol % Y_2O_3)) ceramic material as function of indentation load (P).

P (N)	(c/a)	H (GPa)	$Pc^{-3/2}$ ($MN.m^{-3/2}$)	K_{Ic} ($MPa.m^{1/2}$)	Formula used
49.05	2.00	13.91	64.656	5.09	(1)
				5.32	(2)
				4.85	(3)
				3.26	(4)
				3.76	(5)
				4.93	(6.b)
				6.79	(7)
				4.01	(8)
				4.52	(9)
				6.13	(10)
				7.56	(11)
98.10	2.33	13.01	60.791	4.88	(1)
				5.00	(2)
				5.14	(3)
				3.40	(4)
				3.91	(5)
				5.08	(6.b)
				6.37	(7)
				3.81	(8)
				4.26	(9)
				5.06	(10)
				7.41	(11)
196.30	2.63	12.86	59.719	4.69	(1)
				4.91	(2)
				5.51	(3)
				3.35	(4)
				3.86	(5)
				6.01	(6.a)
				5.42	(6.b)
				6.24	(7)
				3.77	(8)
				4.20	(9)
				4.51	(10)
				7.47	(11)
293.40	2.72	13.02	63.469	4.91	(1)
				5.23	(2)
				5.94	(3)
				3.40	(4)
				3.91	(5)
				6.35	(6.a)
				6.68	(6.b)
				6.57	(7)
				3.81	(8)
				4.28	(9)
				5.06	(10)
				7.97	(11)

CONCLUSIONS, COMMENTS AND RECOMMENDATIONS

The results provided in the preceding sections have shown the problems encountered during the application of Vickers micro-indentation technique and the use of the empirical formulae available in the literature to calculate the fracture toughness of yttria-tetragonal zirconia polycrystals monolithic ceramics and yttria-partially-stabilized zirconia ceramic materials. The following conclusions, comments and recommendations are outlined in this section as follows:

1. Vickers micro-indentation technique is a suitable testing method to measure the fracture toughness of transformation-toughened zirconia-based ceramic materials, however, the phenomenon of stress-induced t->m phase transformation during the indentation process must be thoroughly understood to accommodate the change-of-state that accompanies such a phase transformation. This understanding should provide help to develop a chemithermodynamic model to describe such a fascinating phenomenon.

2. The effects of microstructure, stabilizer's content, grain size, grain size distribution throughout the microstructure and their orientation with respect to crack propagation path must be investigated, since the transformation process is highly dependent on these parameters and, subsequently, crack length measurements and KIc values.

3. The empirical formulae that are available in the literature to calculate the fracture toughness, KIc, of the so-called "well-behaved" brittle materials are NOT appropriate to be used to measure KIc for yttria-tetragonal zirconia polycrystals (Y-TZP) monolithic ceramics using Vickers micro-indentation technique. Therefore, a new model capable of describing the indentation process, taking into account the significant increase in compressive stresses associated with the t -> m transformation process, will be highly required.

4. The new models should be capable of accommodating the parameters mentioned above in point 2, and subsequently, for each class of zirconia-toughened ceramic materials.

5. The new model should be versatile to account for the temperature effects on the fracture toughness of such materials, since the transformability of t -> m phase transformation of $ZrO2$ is a temperature-dependent phenomenon, which will affect the reliability of such materials under high temperature, load bearing structural applications.

ACKNOWLEDGEMENT

The author would like to express his sincere thanks to those who were involved in his research activities over the last two decades and contributed to this paper by comments, reviews and stimulating discussions.

REFERENCES

1. R C Garvie, R H J Hannink and T T Pascoe, Nature (London), 258[5337] pp. 703-704 (1975).
2. D L Porter and A H Heuer, J Am Ceram Soc., 60 [3-4],183-184 (1975).
3. N Claussen, in "Advances in Ceramics", vol. 12, "Science and Technology of Zirconia II," N Claussen, R Ruehle and A H Heuer, Eds., pp. 325-354 (1984).
4. A G Solomah, in Proceedings of the 2nd Int. Conf. on "Ceramic Powder Science and Tecnology," H Hausner and Gl Messing, Eds., (Deutsche Keramische Gesellschaft, DKG, Koeln, Germany) pp. 861-867 (1989).
5. A G Solomah, in Abstracta of the 90th Ann Meet Am Ceram Soc., Cincinnati, OH, May 1-5 (1988).
6. A G Evans and T R Wilshaw, Acta Metall., 24, pp. 939-956 (1976).
7. A G Evans and E A Charles, J Am Ceram Soc., 59[7-8], pp. 371-372 (1976).
8. J E Blendel, PhD Thesis, MIT, Massachausetts, (1979).
9. B R Lawn, A G Evans and D B Marshall, J Am Ceram Soc., 63[9-10] pp. 574-581 (1980).
10. D B Marshall and A G Evans, J Am Ceram Soc., 64, C-182 (1981).
11. G R Anstis, P Chantikul, B R Lawn and D B Marshall, J Am Ceram Soc., 64 [9] pp.533-538 (1981).
12. K Niihara, J R Morena and D P H Hesselman, J Mat Sci Lett., 1, 13-16 (1982).
13. L Lankfoed, J Mat Sci Lett., 1, 493-495 (1982).
14. M T Laugier, J Mat Sci Lett., 4, 1539-1541 (1985).
15. K Tanka, J Mat Sci., 22, 1501-1508 (1987).
16. M T Laugier, J Mat Sci Lett., 6, 355-356 (1987).
17. K M Liang, G Orange and G Fantozzi, J Mat Sci., 25, 207-214 (1990).
18. A G Solomah, C D Cann, R C Styles and K George, in Abstracts Ann Meet Canad Inst Metallurgists (CIM), Winnipeg, MB, August 23-25 (1987).
19. A G Soloamh and W Reichert, in Abstracts the 91st Ann Meet Am Ceram Soc., Indianapolis, IN, April 23-27 (1989).
20. A G Solomah, Unpublished Results (1992).
21. A G Solomah, SAC International Ceramics, TR-07 (1999).

VICKERS INDENTATION: A POWERFUL TOOL FOR THE ANALYSIS OF FATIGUE BEHAVIOR ON GLASS

Vincenzo M. Sglavo and Massimo Bertoldi
Dipartimento di Ingegneria dei Materiali e Tecnologie Industriali
Università di Trento
Via Mesiano 77
38050 Trento, ITALY

ABSTRACT

Indentation undoubtedly represents the most widely used technique for fracture and fatigue characterization of glasses, ceramics and, in general, brittle materials. Especially Vickers indentation cracks have been extensively used for the analysis of sub-critical crack growth behavior in glasses and ceramics. In this work, results regarding sub-critical growth of indentation cracks upon loading in silicate glass are reviewed, analyzed and used to characterize the material in terms of fatigue susceptibility and fatigue limit. Both *normal* and *anomalous* glasses are considered; therefore, sub-critical propagation of both median and cone cracks is studied.

INTRODUCTION

Fatigue is a well-known phenomenon typical for glasses and several ceramic materials.[1-3] In the case of silicate glasses, it consists in a stress-enhanced chemical reaction that takes place between a reactive species, usually water or hydroxyl ion and the stressed siloxane bonds at the crack tip.[4-6] Such interaction leads to sub-critical propagation of cracks, which results in a decrease of the strength with time and in premature failure. The design of glass structural components must proceed through the estimation of the maximum stress, which guarantees a minimum failure time.

The most common relation proposed to define sub-critical crack velocity, v, as a function of the applied stress intensity factor, K_I, is the power-law:[7]

$$\frac{dc}{dt} = v = v_0 \left(\frac{K_I}{K_{IC}} \right)^n , \qquad (1)$$

where n and v_0 are parameters depending on the material and the environment and

K_{IC} is the fracture toughness. The exponent n is also known as the fatigue susceptibility, *i.e.* high values of n correspond to limited fatigue effects. Therefore, lifetime estimate requires the knowledge of the couple (n, v_0) in the specific environment.[2-3]

For several material/environment couples, there are also experimental evidences that v approaches to zero when K_I decreases below a characteristic threshold value, termed fatigue limit, K_{th}, where fatigue no longer occurs.[8-17] From an engineering point of view, the existence of a fatigue limit is extremely desirable in the mechanical design because, below this value, delayed failure does not occur. In spite of the evident importance of this parameter, difficulties have been encountered for its determination, related to the very long time needed to study the low-velocity crack growth range and to the typical large scatter of glass strength data. A certain number of works have been devoted in the past to the determination of K_{th} in glasses and ceramics. In some cases, the fatigue limit was extrapolated from results obtained by static fatigue tests.[18] In other papers, the threshold stress intensity factor is extrapolated from crack velocity measurements.[9,14,19] Nevertheless, both approaches require very long test duration and the identification of a velocity equal to zero is always quite difficult. An alternative technique, based on sub-critical growth of Vickers indentation cracks upon loading, was recently proposed by Sglavo and Green.[20] In the present work such approach is reviewed and applied to characterize different silicate glasses in terms of fatigue susceptibility and fatigue limit.

EXPERIMENTAL PROCEDURE

Silicate glasses of different composition were considered in the present work. Labels, composition and transition temperature, T_g, of the glasses are shown in Table I.

Samples for indentation tests were prepared as bars of nominal dimensions 5-10 x 3 x 80 mm^3, cut from cast glass or from supplied sheets and then polished with SiC papers up to 1200 grit to remove macroscopic defects. In the case of cast glasses (BS2 and BS3), the surface to be indented was also polished with diamond pastes down to 1 μm in order to remove all defects, which could interact with indentation cracks. Great care was used in order to obtain flat and parallel surfaces. Samples were then annealed for 24 h, each glass at about 20°C below T_g, and then cooled down at 2°C/min in order to relax any residual stress. Silica glasses were not annealed.

Indentation tests were performed either in deionized water (corrosive environment) or in silicone oil (inert environment).[21]

Tests in water were performed with duration ranging from 15 s to 6 days using loads of 9.8 N and 39.2 N. The indenter speed was set to about 15 μm/s to avoid any impact damage. A droplet of water was placed on the sample surface before indentation. Additional droplets were added during the test, when needed. All tests were conducted at room temperature on at least 10 samples for each condition.

Tests in silicone oil were conducted on at least 10 samples for each glass, holding the maximum load for 15 s and 150 s. The samples were initially dried at 120°C. A drop of silicone oil was placed on the glass surface before its extraction from the oven, the sample being immediately indented in the region of the oil drop.

Within a minute from test completion, samples were cleaned and broken manually using the indentation crack as critical defect. Sub-surface crack geometry and size were then observed and measured by an optical microscope.

Table I. Composition (wt%) and transition temperature, T_g (°C), of the glasses used in this work

glass	SiO_2	Na_2O	CaO	MgO	K_2O	Al_2O_3	B_2O_3	other	T_g
S1	100								
S2	99.7							0.3	
BS1	80.2	3.5			0.7	2.4	13.2		557
BS2	81.5	9.2					9.3		607
BS3	78.6	13.2					8.2		578
SLS	70.6	12.8	10.5	4.0	0.4	0.8		0.9	574
SAS	62.3	12.8	0.3	3.3	3.5	16.4		1.4	630

RESULTS AND DISCUSSION

Figure 1 represents the typical sub-surface crack systems obtained in the present experimental work. The appearance of cone or median/radial crack was observed to depend on glass composition and test environment.[22]

Conventionally, depending on the plastic deformation mechanism that is active under high shear and pressure stress fields, glasses are usually classified as *normal* or *anomalous*.[23] In normal glasses, deformation occurs mainly by non-

Figure 1. Sub-surface (a) cone crack in silica glass and (b) median crack in soda-lime silicate glass.

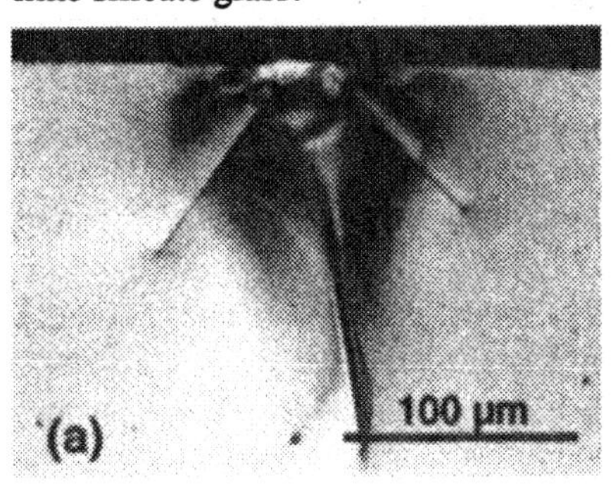

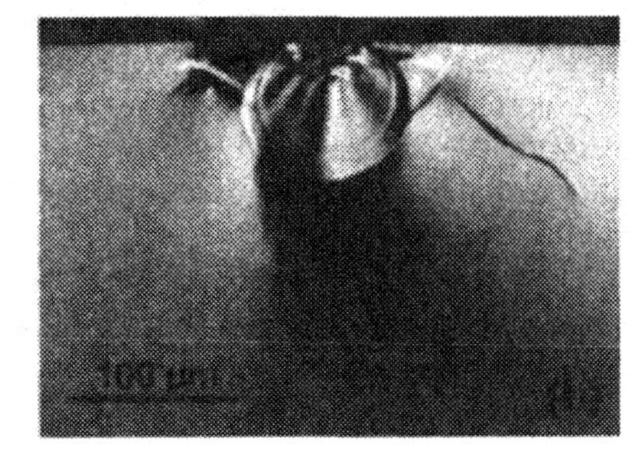

homogeneous shear fault mechanism, whereas in anomalous glasses, inelastic densification due to distortion of silica tetrahedron bonds or to the increase in the average co-ordination of Si atoms represents the main plastic mechanism.[1,23] Such classification is also used to define the indentation behavior, and glasses are usually termed *normal,* when a median/radial crack system is produced, and *anomalous,* when the main crack system produced by a sharp indenter is conical. Quite surprisingly, two of the borosilicate glasses examined in this work showed both behaviours in deionized water as reported in Table II. In any case, cracks of the same system (C or M) obtained in the two different environments were geometrically similar, *i.e.* cone cracks formed constant angles with the sample surface and median cracks were characterized by equal depth to width ratio even regardless the duration of the test and the environment.

Table II. Crack system as a function of test environment (C = cone crack; M = median/radial crack)

glass	deionized water	silicone oil
S1	C	C
S2	C	C
BS1	C	C
BS2	35% C – 65% M	M
BS3	5% C – 95% M	M
SLS	M	M
SAS	M	M

In the case of indentation cracks the relation between stress intensity factor and crack length is usually expressed by the following formula:[1-3]

$$K_I = \frac{\chi P}{c^{3/2}}, \tag{2}$$

where c is the crack length, P the applied load and χ a shape factor depending on the crack system and the direction chosen to measure the crack length. Equation (2) was proved to well-define flaw propagation for both median/radial[1] and conical[24] crack system. During the indentation test, c represents therefore either the cone crack length or the depth of the median crack. Due to the crack geometrical similarity as a function of the environment and test duration previously pointed out, χ can be assumed as a constant for each specific glass and crack system.

The factor χ can be evaluated by Eq. (2) from indentation crack length measured in inert conditions once the fracture toughness of each glass is known. Table III reports values of K_{IC} and χ for the glasses considered in the present work.

Cone crack length and median crack depth are plotted as a function of the indentation dwell-time in Fig. 2. Longer cracks are obtained in water and an asymptotic value is reached for the longest durations. Conversely, defects obtained in silicone oil are shorter and their length does not depend on dwell-time.

Important information on the fatigue behavior of the glass can be obtained by using the data in Fig. 2. Crack length values recorded at times less than 10^4-10^5 s correspond to the sub-critical growth of cracks under a specific stress field. The experimental approach used here allowed the propagation of sub-surface cracks without any interaction with other crack systems such as lateral cracks.[28] Equation (2) can be combined with Eq. (1) to obtain a relation between the crack length and indentation time, t, as:[21]

$$c \cong c_0 \left(h v_0 \frac{t}{c_0} \right)^{1/h}, \tag{3}$$

where c_0 is the crack length in silicone oil and $h = 1.5\,n + 1$. Equation (3) can be easily transformed into a linear relation:

$$\log c = \log \left[c_0 \left(\frac{h v_0}{c_0} \right)^{1/h} \right] + \frac{1}{h} \log t. \tag{4}$$

The experimental data shown in Fig. 2 for $t < 10^4$ s were expressed in terms of log c *vs.* log t and linear regression was used to determine n and v_0. The results calculated for the different glasses are reported in Table III. Obtained data are substantially in agreement with n and v_0 values previously reported for silicate glasses.[1-3] More in detail, the calculated n values for pure silica (S1), borosilicate glasses (BS1, BS2 and BS3) are coincident to typical literature data. The same can be said also for soda-lime-silica (SLS) and aluminosilicate (SAS)

Table III. Fracture toughness (K_C), shape factor (χ), fatigue parameters (n, v_0) and fatigue limit (K_{th}) of examined glasses

glass	K_{IC} (MPa $m^{0.5}$)	χ	n	v_0 (m/s)	K_{th} (MPa $m^{0.5}$)
S1	0.77[27]	0.062	36±6	16.4×10^{-3}	0.35
S2	0.75[1]	0.044	21±4	3.9×10^{-3}	0.28
BS1	0.88[22]	0.042	37±2	5.8×10^{6}	0.32
BS2	0.88[22]	0.033	20±3	2.4×10^{3}	0.20
BS3	0.90[22]	0.030	17±2	1.3×10^{3}	0.18
SLS	0.72[28]	0.026	26±7	3.7×10^{4}	0.21
SAS	0.79[29]	0.026	26±7	3.5×10^{5}	0.21

Figure 2. Crack length as a function of loading time. Empty symbols represent crack length obtained in silicone oil. Indentation load and crack system are also reported.[20,22,24]

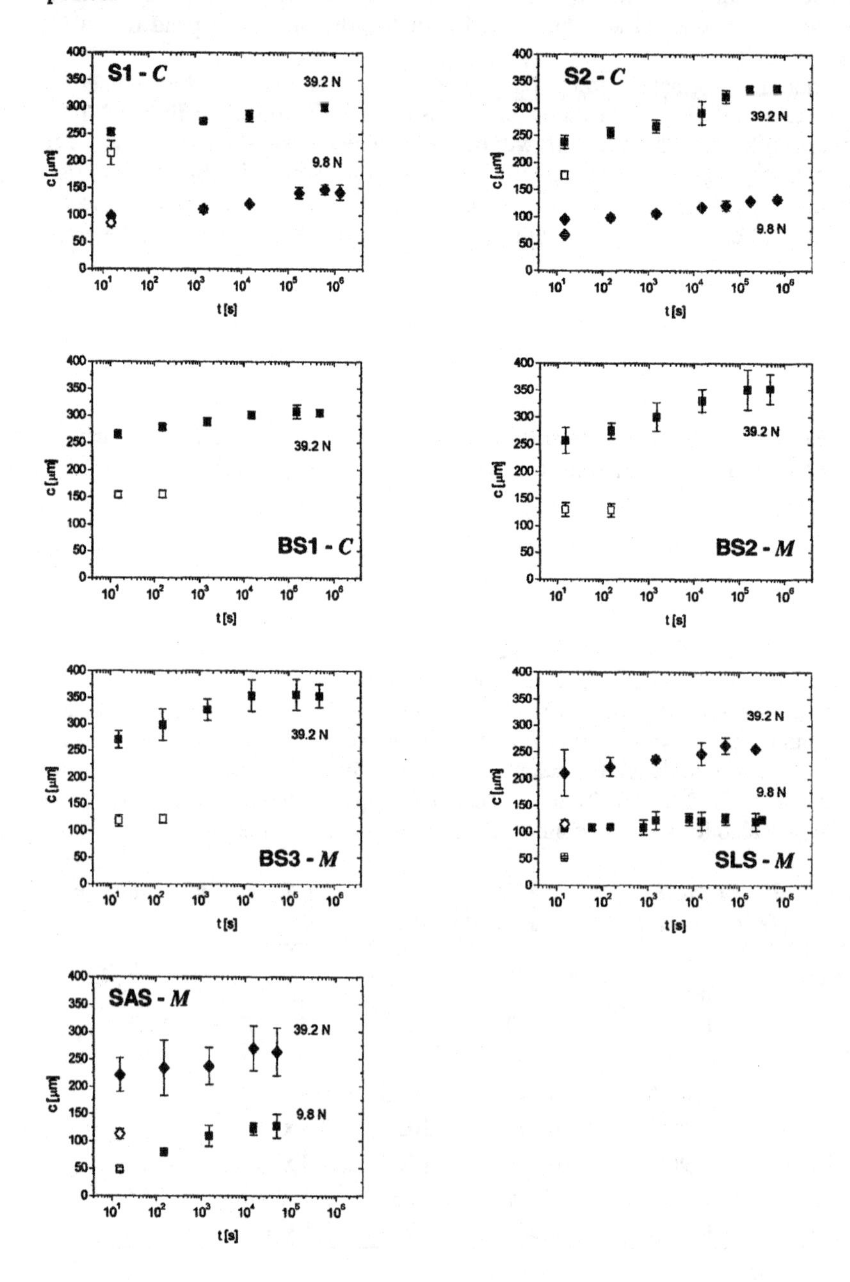

Figure 3. Stress intensity factor applied to indentation cracks. Indentation load is also reported.

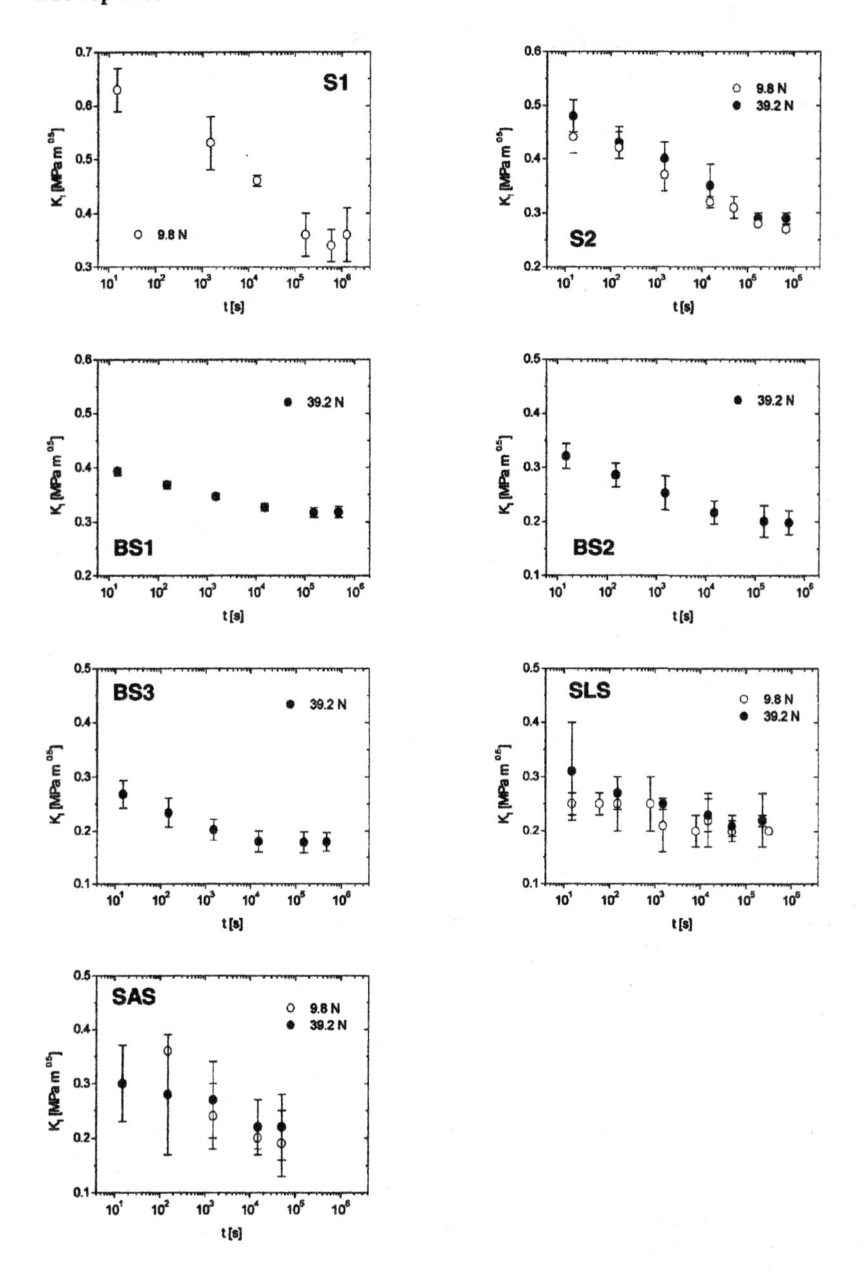

glasses in spite of the large scatter. Only the n values evaluated for the impure silica (S2) can be considered somehow peculiar as it differs substantially from the values calculated for S1 glass. One can observe that the two glasses differ substantially also in the factor χ, related to the elastic modulus/hardness ratio[1] and, therefore, to the strength of the glass structure. Values of n as low as 14 have been obtained in the past by static or dynamic fatigue tests on silica fibers in humid environment.[25,26] In addition, it is not rare to find lower values for n when the lower part of the $v(K)$ curve is explored.[19,27] On the basis of the results obtained here, the hypothesis can be that, even limited amount of impurities have a strong influence on glass structure and fatigue behavior, especially in the lower part of the $v(K)$ curve.

The comparison of the v_0 values calculated in the present work with previous works appears more difficult. In most papers dealing with glass fatigue, v_0 values are not reported and usually the scatter in the results is very large. In fact, the calculation of such parameter uses an evaluation of a power (10^x or e^x), where small differences in the argument can develop into large variations in the final result.

The stress intensity factor applied to the cracks reported in Fig. 2 was evaluated by using Eq. (2). The results are shown in Fig. 3. For time durations longer than 100 ks, an invariant stress intensity factor is reached for all glasses. This allows one to calculate the fatigue limit, K_{th}, reported in Table III. Also in this case, obtained results agree well with previous data of threshold for fatigue on silicate glasses.[9,12,13,15-17,19]

A strong correlation between glass composition and K_{th} values can be observed. Lower fatigue limit values are obtained for higher alkali content (Na_2O + K_2O) in the glass composition. Therefore, K_{th} scales with the hydrolytic resistance of the glass and can be related to the presence of non-bridging oxygens in the glass structure depending on alkali and alkaline earth, B_2O_3 and Al_2O_3 content.

CONCLUSIONS

An experimental approach based on the observation of the sub-critical growth on indentation cracks (median flaws and cone cracks) at maximum load is used for the analysis of fatigue behavior of glasses. The proposed method represents an easy technique that requires small amount of material and whose results are simple to be analysed. In addition, the sub-critical growth of the cracks can be directly observed and tests duration is usually limited.

The proposed method has been used on different silicate glasses to evaluate both fatigue parameters (n, v_0) and fatigue limit (K_{th}) values in humid environment. In most cases, results in good agreement with previous data have been obtained. For pure silica glass values of n and K_{th} equal to 36 and 0.35 MPa $m^{0.5}$, respectively, have been obtained. Such values are strongly dependent on the presence of impurities as $n = 21$ and $K_{th} = 0.28$ MPa $m^{0.5}$ are measured for 99.5 wt% pure silica. Similar arguments have been advanced for borosilicate glasses:

n changes from 37 to 17 and K_{th} from 0.32 to 0.18 MPa $m^{0.5}$, when the alkali content (Na_2O + K_2O) is increased from 4.2 to 13.2 wt%. For soda-lime and soda-alumina silicate glasses the obtained results are $n \approx 26$ and $K_{th} \approx 0.21$ MPa $m^{0.5}$, again well comparing with previous data.

The experimental technique investigated here can also be applied to other ceramic and glass-ceramics materials in different environmental conditions, once the experimental apparatus has been specifically set up.

REFERENCES

[1]B. R. Lawn, *Fracture of Brittle Solids: Second Edition*, Cambridge University Press, Cambridge, UK, 1993.

[2]D. J. Green, *Introduction to Mechanical Properties of Ceramics*, Cambridge University Press, Cambridge, UK, 1998.

[3]D. Munz and T. Fett, *Ceramics. Mechanical Properties, Failure Behaviour, Materials Selection*, Springer-Verlag, Heidelberg, Germany, 1999.

[4]S. M. Wiederhorn, "A Chemical Interpretation of Static fatigue", *J. Am. Ceram. Soc.*, **55** [2] 81-85 (1972).

[5]T. A. Michalske and S.W. Freiman, "A Molecular Mechanism for Stress Corrosion in Vitreous Silica", *J. Am. Ceram. Soc.*, **66** [4] 284-288 (1983).

[6]T. A. Michalske and B. C. Bunker, "Steric effect in Stress Corrosion Fracture of Glass", *J. Am. Ceram. Soc.*, **76** [10] 2613-2618 (1993).

[7]A. G. Evans, "A Method for Evaluating the Time-Dependent Failure Characteristics of Brittle Materials and Its Application to Polycrystalline Allumina", *J. Mater. Sci.*, **7** 1137-46 (1972).

[8]W. B. Hillig, and R. J. Charles, "Surfaces, Stresses-Dependent Surface Reactions and Strength". In *High Strength Materials*, Ed. by V.F. Zakay, John Wiley and Sons, New York, pp. 683-705, 1965.

[9]S. M. Wiederhorn and L. H. Bolz, "Stress Corrosion and Static Fatigue of Glass", *J. Am. Ceram. Soc.*, **53** [10] 543-48 (1970).

[10]A.G. Evans, "A Simple Method for Evaluating Slow Crack Growth in Brittle Materials", *Int. J. Fract.* **9** [3] 267-75 (1973).

[11]B. J. S. Wilkins and R. Dutton, "Static Fatigue Limit with Particular Reference to Glass", *J. Am. Ceram. Soc.*, **59** [3-4] 108-12 (1976).

[12]T. A. Michalske, "The Stress Corrosion Limit: Its Measurement and Application". In *Fracture Mechanics of Ceramics*, Vol. 5, Ed. R. C. Bradt et al., Plenum Press, New York, pp. 277-89, 1983.

[13]R. F. Cook, "Influence of Crack Velocity Thresholds on Stabilized Non-Equilibrium Fracture", *J. Appl. Phys.*, **65** 1902-10 (1989).

[14]K. Wan, S. Lathabai, and B.R. Lawn, "Crack Velocity Functions and Threshold in Brittle Solids", *J. Eur. Ceram. Soc.*, **6** 259-68 (1990).

[15]R. F. Cook and E. G. Liniger, "Kinetics of Indentation Cracking in Glass", *J. Am. Ceram. Soc.*, **76** 1096-1106 (1993).

[16]V. M. Sglavo and D. J. Green, "Threshold Stress Intensity Factor for Soda Lime Silicate Glass by Interrupted Static Fatigue Test", *J. Eur. Ceram. Soc.,* **16** 645-51 (1996).

[17]V. M. Sglavo and S. Renzi, Fatigue Limit in Borosilicate Glass by Interrupted Static Fatigue Test, *Phys. Chem. Glasses,* **40** [2] 79-84 (1999).

[18]R. E. Mould and R. D. Southwick, "Strength and Static Fatigue of Abraded Glass under Controlled Ambient Conditions: Effect of Various Abrasions and the Universal Fatigue Curve", *J. Am. Ceram. Soc.*, **42** [12] 582-592 (1959).

[19]C. Kocer and R. E. Colins, "Measurement of very Slow crack Growth in Glass", *J. Am. Ceram. Soc.*, **84** [11] 1585-93 (2001).

[20]V. M. Sglavo and D. J. Green, "Indentation Determination of Fatigue Limits in Silicate Glasses", *J. Am. Ceram. Soc.,* **82** [5] 1269-74 (1999).

[21]V. M. Sglavo and D. J. Green, "Subcritical Growth of Indentation Median Cracks in Soda-Lime-Silica Glass", *J. Am. Ceram. Soc.*, **78** [3] 650-56 (1995).

[22]M. Bertoldi and V. M. Sglavo, "Influence of Composition on Fatigue Behavior and Threshold Stress Intensity Factor of Borosilicate Glasses", *J. Am. Ceram. Soc.*, **85** [10] 2499- 506 (2002).

[23]A. Arora, D. B. Marshall and B. R. Lawn, "Indentation Deformation/ Fracture of Normal and Anomalous Glasses", *J. Non-Cryst. Solids,* **31** 415-28 (1979).

[24]V. M. Sglavo and D. J. Green, "Fatigue Limit in Fused Silica", *J. Eur. Ceram. Soc.*, **21** 561-67 (2001).

[25]T. A. Michalske, W. L. Smith and B. C. Bunker, "Fatigue Mechanisms in High-Strength Silica-Glass Fibers", *J. Am. Ceram. Soc.*, **74** [8] 1993-96 (1991).

[26]M. Muraoka and H. Abe, "Subcritical Crack Growth in Silica Optical Fibers in a Wide Range of Velocities", *J. Am. Ceram. Soc.*, **84** [5] 51-57 (1996).

[27]R. F. Cook, "Crack Propagation Thresholds: a Measure of Surface Energy", *J. Mater. Sci.,* **1** 852-56 (1986).

[28]V. M. Sglavo and D. J. Green, "Influence of Indentation Crack Configuration on Strength and Fatigue Behaviour of Soda-Lime Silicate Glass", *Acta Metall. Mater.*, **43** [3] 965-72 (1995).

[29]P. J. Dwivedi and D. J. Green, "Determination of Subcritical Crack Growth Parameters by In Situ Observation of Indentation Cracks", *J. Am. Ceram. Soc.*, **78** [8] 2122-28 (1995).

NANO-INDENTATION AS A COMPLIMENTARY METHOD TO XRD AND RCA FOR CALCULATING RESIDUAL STRESS IN THIN FILMS

W. D. Nothwang, M. W. Cole, C. Hubbard, and E. Ngo
United States Army Research Laboratory, WMRD
APG, MD 21005

Residual stress in thin film materials seriously alters the application performance envelope, and depending upon the direction of the stress, it can be either extremely deleterious or advantageous. To more fully understand how residual stress affects the properties and performance of a thin film material for a particular application, it is necessary to first measure the value of the stress. There currently exists no method for directly measuring the residual stress of a material that is both accurate and does not require significant sample preparation. However, there are a number of ways to measure the effects of stress on other mechanical properties, and the stress can be estimated from these measurements. In thin films materials, the two methods used for measuring the strain are: 1) nano-indentation, 2) and, reflective curvature analysis,. In this paper, residual stress measurements are made on metal-organic solution deposited magnesium doped barium strontium titanate thin films on magnesium oxide substrates, indicating that the residual stress within these films is dependent on the annealing temperature and to a lesser degree the doping level.

INTRODUCTION

The impact of residual stress on thin film materials cannot be understated. As a result of decreasing films thickness, the residual stress within the film increases. For piezoelectric films the extrinsic contributions, which account for over 90% of the total bulk piezoelectric response, are not observed due to clamping of the residual stress. The dielectric constant of piezoelectric materials may be decreased by over 50%, and the dielectric loss is also significantly increased as electron tunneling and channeling become much more prevalent. Stress also accelerates the fatigue of these devices by several orders of magnitude, such that at elevated stress levels a device with a standard lifetime of 10 years under normal conditions would degrade in under 30 days. Stress also significantly affects the mechanical properties of thin film materials. For example, residual stresses often result in cracks, fracture, and failure during normal in service device operation. Traditionally, these deleterious effects have been attributed to elevated tensile stresses within the films, and it has been theorized that compressive stresses will improve the properties within the films.

Residual tensile stress causes the measured hardness to be less than the stress free hardness, and compressive residual stress causes the measured hardness to exceed the zero stress value. Films with a highly compressive stress will have a much higher hardness than films with no stress, and similarly the modulus will be much higher for compressively stressed films than films in tension.[10] The fracture toughness and modulus of rupture will be strongly influenced by the magnitude and sign of the residual stress.

By varying the deposition conditions, it is possible to affect the residual stress within a device, and necessarily the mechanical properties. Jong et al. observed a strong relationship between deposition conditions and the residual stress level during their studies of RF-sputtered TaON thin films, and correlated the stress to the density of the developed film.[9] Dauskardt et al. examined film debonding, which is strongly dependent upon the level of residual stress.[6] The crack propagation rate drastically increases as the residual stress increases. Most importantly they noted that the mechanical lifetime of thin films bonded to silicon substrates decreases with increasing stress levels.[6] The effects of residual stress will also be significant for thin film fatigue, fracture, wear, corrosion, and friction.[3,10] Residual stress within a film will broaden many transitions observed in the bulk and make them less intense. Some of the transitions affected are Curie transition,[17] phase transition,[13] elastic to plastic strain transition.[5] Shaw et al. suggested that the thin film under high residual stress conditions might undergo a phase transformation from the paraelectric phase to a ferroelectric phase with polarization inline with the stress, which was independently corroborated.[13,17]

The residual stress is of particular importance for the dielectric properties of thin films for voltage tunable device applications. For such thin film dielectrics, an electrical field of the order of 2 V/μm is typically applied. This electrical field can induce significant stress in a film that may otherwise be nearly stress free. Shaw et al. observed that a residual tensile stress of 600 MPa in a thin film of barium strontium titanate (BST) caused a decrease of 23% in the dielectric constant (44 $fF/\mu m^2$ vs. 35 $fF/\mu m^2$).[17] In BST thin films, Yamamichi et al. observed a severe increase in leakage current as the electrical field increased, from $5*10^{-8}$ at 0 C/cm^2 to $3*10^{-6}$ at 6.7 C/cm^2, which was attributed to the electrical field induced stress.[19] They also observed that thinner films (50 nm), which have a higher level of residual stress, experienced an earlier onset of leakage current than thicker films (160 nm). The leakage current for the 50 nm films can be up to 100 times greater than the leakage current for the 160 nm films. This increase in leakage current drastically affects the dielectric reliability and/or lifetime of the thin film device. A film with minimal stress has a dielectric lifetime of ~10 years, while a film with several hundred MPa of tensile stress has a lifetime of only a few weeks. Yamamichi et al. also noted that the dielectric constant for the 50 nm film was 240, while for the 160 nm film it was 450.[19] Such an increase was quantified by Poulane et al..[14] For 2μm thick films, there was a residual stress of 25 MPa, while for a thickness of 5.5 μm, the residual stress was 12 MPa.[14] This increased stress, which can be up to three orders of magnitude greater in ceramic films, has drastic effects on the mechanical, dielectric, and fatigue behavior of thin films.

Residual stress acts as a driving force for many defect mechanisms, and the incorporation and/or migration of the defects affect the residual stress in the film. If monitored over time, these results can be easily observed and quantified. Over a time period of six days for TiO_2 and MgF_2 films, Atanassov et al. noted that the stress decreased significantly, and the rate of decrease depended heavily on the deposition technique.[1] Using plasma assisted ion deposition, Atanassov was able to reduce the initial residual stress, and to increase the rate of stress relaxation as compared to traditional electron beam evaporation. The difference is explained using grain boundary models, where they hypothesize that lattice relaxation forced by the energetic interaction between neighboring grains can be observed as a mechanical elastic deformation. The ion bombardment forces a densification of the films (and grain structure), and this leads to a decrease in residual stress. If grains are brought close enough together, a compressive stress may result.[1]

Residual stress in thin films can arise due to several different factors. Most prominent among these are film condensation during deposition, lattice mismatch, differences in thermal expansion coefficient, defect inclusions, and dielectric/piezoelectric displacement. It is often difficult to distinguish between each, and traditionally the effects are treated additively.[17] In this relationship, σ_p is the stress due to the condensation of the film; Δd_l is the lattice mismatch; M_f is the biaxial modulus, which is a function of the Young's Modulus (E_f) and the Poisson's ratio (v_f) of the material; $\Delta\alpha$ is the difference in thermal expansion coefficients; T is the temperature; and, Q_{12} and D are the electrostrictive coefficient ($Q_{12} < 0$) and the electric displacement, respectively, which result in the electrostrictive strain.

$$\sigma = \sigma_P + M_f\left(\Delta d_l + \Delta\alpha\Delta T - Q_{12}D^2\right),$$
$$M_f = \frac{E_f}{\left(1 - v_f\right)}. \qquad \text{Eq 1}$$

Nano-indentation operates by measuring the force required to deform a material with a diamond indenter. There are three types of indenter heads used in nano-indentation systems. A Vickers head is a sharp pyramidal shaped indenter; a Berkovich head is a sharp three-sided tip, as opposed to the four-sided Vickers head; while, a Hertzian indenter is rounded. The blunt tips differ drastically from the sharp tips in the manner of the deformation that is induced. The blunt tips deform elastically up to the point of fracture, while sharp tips are dominated by plastic deformation.[12]

Nano-indentation offers a means of indirectly estimating the residual stress in a thin film, when using a sharp tip, by measuring the hardness of the film and substrate. Compressive stresses increase the apparent hardness of a film, while tensile stresses decrease the apparent hardness. These effects are difficult to measure, however, unless the stress level is at least several hundred MPa.[13]

The physical approximations and predictions for the effects of residual stress on hardness and modulus are based on several finite element analysis studies. Giannakopoulos and Suresh were able to predict the effects of an indenter on an anisotropic, non-homogeneous material system by allowing for the effects of residual stress and complex modulus.[18] Daehn et al. used a similar method to examine composite effects on Young's modulus.[5] To calculate the Young's modulus of a sample a force versus depth curve is measured using a nano-indentation instrument during increasing and decreasing load, as shown in Figure1. From this curve the Young's modulus and hardness can be obtained by:

$$\frac{dh}{dP} = \left(\frac{\pi}{4A_P}\right)^{1/2} \frac{1}{E_R}, \qquad \text{Eq 2}$$

$$\frac{1}{E_R} = \frac{1-\nu^2}{E} + \frac{1-\nu_0^2}{E_0}, \text{ and} \qquad \text{Eq 3}$$

$$H_V = \frac{P_{max}}{A_P}. \qquad \text{Eq 4}$$

The slope of the unloading curve, or the compliance, is given by dP/dh, and it is frequently referred to as the stiffness (S). The contact depth, h_P, is the plastic depth of the indent, and the terms are used interchangeably. E_R is the measured modulus of the system, and it is related to the true Young's modulus (E) of the film. The indenter characteristics are represented by its Young's modulus (E_0) and Poisson's ratio (ν_0), and the Poisson's ratio of the sample is ν. H_V is the Vicker's hardness, and the maximum applied load is P_{max}.[8] For a Berkovich tip A_P is 24.5 h^2_P; the cross sectional area for a Vickers tip, A_P, is 26.429h^2_P; and, for a conical tip it is $(2\pi/\tan^2\alpha)\,h^2_P$, where α is the conical angle.[7]

The modulus, as well as other material properties, is directly affected by the presence of residual stress within a material. Because of this, residual stress can be calculated from the data collected in a hardness test if specific material properties are known. While all materials possess an inherent residual stress level, it is generally assumed that the bulk materials are stress free. There are two methods of calculating residual stress. If bulk values for modulus and poison's ratio are available, they are used as a reference value; otherwise, an internal reference is used. An in-depth analysis of the two methods is given by Suresh and Giannakopoulos.[18]

For both methods, first a *P-h* curve must be determined, and in addition to the maximum penetration depth (h_{max}), maximum pressure (P_{max}), and unloading rate (dP/dh), it is also necessary to record the plastic depth (h_p), the elastic depth (h_e), and the complete unloading depth (h_r). These are easily determined from the *P-h* curve, as shown in Figure 1.[18]

$$h_r + h_e = h_{max},$$
$$h_p + h_e/2 = h_{max}. \qquad \text{Eq 5}$$

Next, it is necessary to experimentally determine the contact area. Typically this can be done with AFM or SEM; however, if the properties of the indenter are very well understood, or it is not practical to do AFM/SEM measurements, the contact area may be approximated. Once the area has been found, the unloading curve can be determined doing a traditional pressure at depth, or depth at pressure measurement. From the data collected, it is possible to estimate the stress regime. If $P_{max}/(A_{max}E\tan\alpha) \leq 0.1$, then it is possible to calculate the stress directly assuming that the uniaxial residual stress (σ_u) is equal to the stress in the y direction (σ_y) is equal to the residual stress in the y direction (σ_y^R):

$$\sigma_u \approx \sigma_y \approx \sigma_y^R \approx \frac{P_{max}}{2.8A_{max}}; \qquad \text{Eq 6}$$

however, $P_{max}/(A_{max}E\tan\alpha) > 0.1$ a set of simultaneous equations must be solved to estimate the residual stress:, which are rearranged into the following form:

$$\sigma_y^R = -0.145E\left(1 - 0.1419\left(\frac{h_r}{h_{max}}\right) - 0.9568\left(\frac{h_r}{h_{max}}\right)^2\right) + \left(\frac{P_{max}(\tan\alpha)^2}{2c^*h_{max}^2}\right)\left(\frac{1}{1+\ln\left(\frac{E\tan\alpha}{3\sigma_y^R}\right)}\right)$$

Eq 7

and c^* is 1.220 for a Vicker's tip, and 1.273 for a Berkovich. The angle at the tip of the indenter is represented by α, and it is:

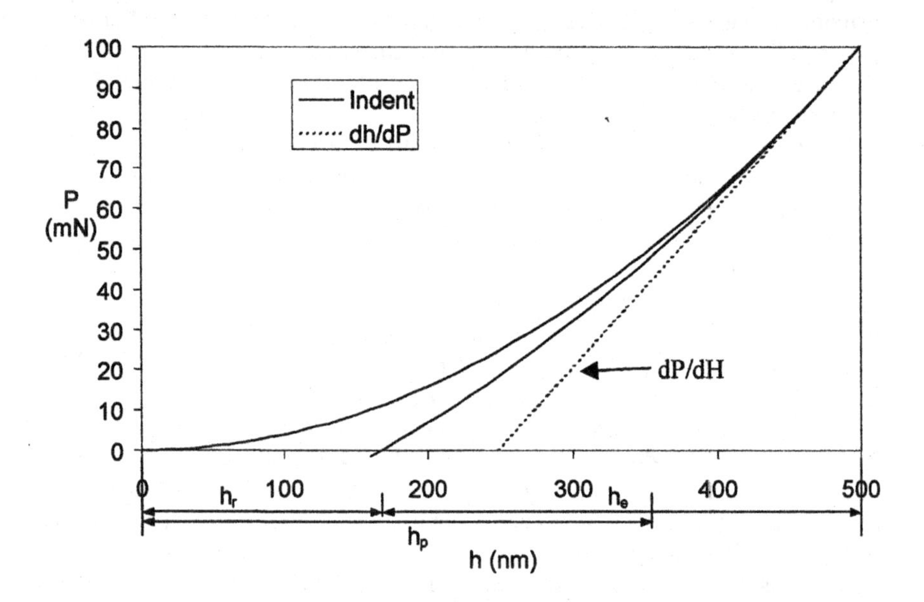

Figure 1: A representative loading and unloading curve can be used to calculate the residual depth (h_r), the elastic depth (h_e), and the plastic depth (h_p).

$\alpha = 22^\circ$ for Vickers,

$\alpha = 24.7^\circ$ for Berkovich, and

$\alpha = 19.7^\circ$ for Equivalent Circular Conical.

It is easily solved using the initial approximation of:

$$\left(\sigma_y^R\right)_{I.G.} = \frac{P_{\max}}{2.8A_{\max}}. \qquad \text{Eq 8}$$

If the material conditions outlined in Equation 9 are satisfied, it is possible to simplify Equation 7. If the value of Equation 9 is less than 0.1, then the initial estimate shown in Equation 8 can be used as a reasonable approximation.

$$P_{\max}/\left(A_{\max}E\tan\alpha\right) \leq 0.1. \qquad \text{Eq 9}$$

A far simpler method is to measure the radius of curvature of the sample before and after the deposition of the film. Examining the change in the radius of curvature caused by the strain within the film it is possible to estimate the extent of stress in the film. Again, it is necessary to know some of the material parameters of the film and substrate and to make some assumptions. It is necessary to assume that there is an equibiaxial stress present, and that any changes in the radius of curvature are caused solely by the presence of the film. With these assumptions, the Stoney equation can be used to calculate the amount of total stress in the system:

$$\sigma_{total} = \frac{E_S h^2}{(1-\nu_S)6d}\left(\frac{1}{R_f} - \frac{1}{R_i}\right), \qquad \text{Eq 10}$$

where, the radius of curvature is R. The substrate Modulus (E_S) and Poisson's Ratio (ν), together with the substrate thickness (h) and the film thickness (d) combine to render the total film stress (σ_{total}).[1]

EXPERIMENTAL PROCEDURE

Magnesium doped BST thin films were fabricated by metalorganic solution deposition (MOSD) on single crystal (100) magnesium oxide The samples were spin coated with 5 coats to achieve a nomial, post-anneal thickness of 190 nm. The films were pyrolyzed at 350°C between each coat for ten minutes and 400°C for 30 minutes following the final coat. Films were annealed in flowing oxygen at temperatures between 650 and 950°C.

Residual stress was measured using two different techniques. 1) Reflective curvature analysis (RCA) was done using a Flexus 2100 system. An automated scan was conducted using two laser sources (670nm and 750 nm) for optimum results. Substrates were scanned before film deposition, after pyrolization of the final coat, and after annealing. System reference valuse for the material and orientation of the substrates were used. 2) Indentation was performed with a NanoIndentor v1.1. Pressure to depth experiments were conducted using a new Berkovich indenter. Prior to indenting the thin films, calibrations were performed on two different standards and a system calibration was performed. All observed errors were within reported machine error (5%). A total of 25 indentations were conducted at target depths of 100, 150, 200, 300 and 400nm for statistical significance. Indentations and film surfaces were examined via AFM, optical and scanning electron microscopy. The indentations areas corresponded to those calculated by the instrument.

RESULTS AND DISCUSSION

The residual stress within the 10% magnesium doped barium strontium titanate samples were tested by RCA (Figure 3 A) and nano-indentation (Figure 2 A) to examine the different components of residual stress within thc materials. The nano-indentation results, shown in Figure 2(A), indicate a gradual decrease in the residual stress with increasing annealing temperature. Nano-indentation measurements for 10 mol% magnesium doped BST up to the film interface (200nm) and into the substrate. Only the 700°C, 850°C, and the 950°C samples are shown for clarity. The RCA measurements capture a similar trend in the data; however, there is a sharp break in the curve between 800°C and 850°C that was not observed in the nano-indentation results. The XRD results, not shown, indicated that at annealing temperatures greater than 800°C the films condensed a second phase (MgO), while below 800°C the material was a single phase crystalline material. It has been reported that in varactor materials of similar composition that a high dopant concentrations forms near the surface of the grains.[20] Above a certain concentration, and at a certain temperature, there is enough free energy available within this surface region to form a second phase. The atomic force microscopy and plan view FESEM/EDS indicated that the grainsize was significantly smaller than the grainsize in pure BST films as well as the nano-indenter tip. The relative magnitude of the residual stress is significantly different between the two techniques, as the RCA measures the composite residual stress and nano-indentation only measure the local stress field. The substrate typically has a large affect on the film properties at the interface, and the strain relaxes throughout the film. As the slope through the film-substrate interface (190-220nm) is almost zero, it can indicate 1) that the films is dominated by substrate effects throughout a significant thickness of the film, 2) or, that the films is completely independent of the substrate effects. A substrate dominated film is much more likely due to the observed increase in stress levels approaching the interface by nano-indentation, and the high total residual stress measured by RCA.

For 5 mol% magnesium doped barium strontium titanate films the residual stress was measured using nano-indentation, as shown in Figure 2(B), and reflective curvature analysis, as shown in Figure 3(B). For clarity only the 800°C, 850°C, and the 950°C samples are shown in Figure 2(B). Both techniques capture the sharp change in residual stress between 800°C and 850°C, as the films become significantly more tensile. The residual stress within the film decreased (i.e. becoming more compressive) up to 800°C, and it demonstrated a large interface region, as observed in the 10 mol% samples, but the stress being felt by the substrate was drastically increasing, as observed by the RCA. Above 800°C a dramatic change in the crystallization behavior of the film was affected. There are several possible explanations for this. It is possible that at these temperatures the kinetics of the nucleation and growth crystallization observed were not fast enough to mitigate the increasing residual stress. Once the residual stress reached a threshold level, a second growth mechanism, the bulk condensation, was triggered. This would results in an interface that was more highly graded interface, as observed at 850 and 950°C, a large number of low stress grains, seen in the circled area, and a large decrease the net compressive stress, as the new grains rapidly form. The AFM results, not shown, also indicate a large increase in grain size between 800 and 850°C. Other possible solutions include, oxygen vacancy considerations, slip plane inclusions, meta-stable phases, and second order kinetic crystallization.

Each method addressed has certain advantages and limitations to measuring the stress. As such, it is necessary to examine what resources are available, and which results are more important. None of the methods is inherently more accurate or superior. As each method observes a different facet of the same property, it is also difficult to compare the results between the methods. Shaw et al. using curvature analysis to estimate the residual stress in BST thin films as they were gradually ablated via ion milling, observed that the stress level was constant through the depth.[17] Saha and Nix, however, using nano-indentation to measure the hardness of aluminum and tungsten films observed that the stress was strongly influenced by the depth.[15] This elucidates one of the limitations of the two different techniques used. Curvature analysis is extremely useful for measuring the total stress within a film, but it can not discern the local stress fields within the film, while nano-indentation can discern the stress at each level of the film, and only by integrating the curve generated, can the total stress present be determined. Coupling this information with the theoretical predictions, outlined above, it is possible to examine the stress relief mechanisms.

These techniques and the theoretical predictions are very complimentary. Reflective curvature analysis can be used to map the stress across a surface with time and temperature, and the stress through the depth can be acquired from the nano-indentation data. Both nano-indentation methods can be used, as the internal reference methods will yield a more accurate result, and the bulk reference method can be employed to insure that the results of the internal reference are reasonable. XRD can be used to verify the bulk stress value of the reflective curvature analysis, and yield information about the crystallite orientation. Crystallite orientation will help to discern how the stress manifests itself within the lattice structure of the film. Combining these analytical results with the theoretical calculations, it is possible gain a more coherent picture of the stress within the thin film.

BIBLIOGRAPHY

1. G. Atanassov, J. Turlo, J.K. Fu, and Y.S. Dai, "Mechanical, optical and structural properties of TiO_2 and MgF_2 thin films deposited by plasma ion assisted deposition," *Thin Solid Films*. **342**, 83-92 (1999).

2. Y. Bisrat and S.G. Roberts, "Residual stress measurement by Hertzian indentation," *Material Science and Engineering.* **A288**, 148-153 (2000).
3. S. Carlsson and P.L. Larsson, "On the determination of residual stress and strain fields by sharp indentation testing. : Part I: theoretical and numerical analysis," *Acta Materialia.*, **49**, 2179-2191 (2001).
4. B.D. Cullity, Elements of X-Ray Diffraction. 2nd Edition, Addison-Wesley Publishing Company, Inc. London, 1978.
5. G.S. Daehn, B. Starck, L. Xu, K.F. Elfishawy, J. Ringalda, and H.L. Fraser, "Elastic and plastic behavior of a co-continuous alumina/aluminum composite," *Acta Materialia.* **44** (1), 249-261 (1996).
6. R.H. Dauskardt, M. Lane, Q. Ma, and N. Krishna, "Adhesion and debonding of multi-layer thin film structures," *Engineering Fracture Mechanics.* **61**, 141-162 (1998).
7. A.C. Fischer-Cripps, "A review of analysis methods for sub-micron indentation testing," *Vacuum***58**, 569-585 (2000).
8. M. Hoffmann and R. Birringer, "Elastic and plastic behavior of submicrometer-sized polycrystalline NiAl," *Acta Materialia.* **44** (7), 2729-2736 (1996).
9. C.A. Jong, T.S. Chin, and W. Fang, "Residual stress and thermal expansion behavior of TaO_xN_y films by the micro-cantilever method," *Thin Solid Films.* **401**, 291-297 (2001).
10. L. Karlsson, L. Hultman, and J.E. Sundgren, "Influence of residual stresses on the mechanical properties of TiC_xN_{1-x} (x=0, 0.15, 0.45) thin films deposited by arc evaporation," *Thin Solid Films.* **371**, 167-177 (2000).
11. S.G. Malhotra, Z.U. Rek, S. M. Yalisove, and J.C. Bilello, "Analysis of Thin Film Stress Measurement Techniques." *Thin Solid Films,* **301**, 45-54 (1997).
12. W.C. Oliver, "Progress in the Development of a Mechanical Properties Microbe," *MRS Bulletin.* B11B(5), 15-19 (9-1986).
13. G.M. Pharr and W.C. Oliver, "Measurement of Thin Film Mechanical Properties Using Nano-indentation," *MRS Bulletin.* **17** (7), 28-33 (7-1992).
14. C. Poilane, P. Delobelle, L. Bornier, P. Mounaix, X. Melique, and D. Lippens, "Determination of the mechanical properties of thin polyimide films deposited on a GaAs substrate by bulging and nano-indentation tests," *Materials Science and Engineering.* **A262**, 101-106 (1999).
15. R. Saha and W.D. Nix, "Effects of the substrate on the determination of thin film mechanical properties by nano-indentation," *Acta Materialia.* **50**, 23-38 (2002).
16. U. Schmid, M. Eickhoff, H. Richter, G. Kroetz, D. Schmidtt-Landsiedel, "Etching characteristics and mechanical properties of a-SiC:H thin films," *Sensors and Actuators.* **A94**, 87-94 (2001).

17. T.M. Shaw, Z. Suo, M. Huang, E. Liniger, R.B. Laibowitz, and J.D. Baniecki, "The effect of stress on the dielectric properties of barium strontium titanate thin films," *Applied Physics Letters.* **75** (14), 2129-2131 (10-4-1999).
18. S. Suresh and A.E. Giannakopoulos, "A new method for estimating residual stresses by instrumented sharp indentation," *Acta Materialia.* **46** (16), 5755-5767 (1998).
19. S. Yamamichi, A. Yamamichi, D. Park, T.J. King, and C. Hu, "Impact of time dependent dielectric breakdown and stress induced leakage current on the reliability of high dielectric constant (Ba,Sr)TiO_3 thin film capacitors for Gbit-scale DRAMs," *IEEE Transactions on Electron Devices.* **46** (2), 342-347 (2-1999).
20. R. Buchanan, **Ceramic Materials for Electronics: Processing, Properties and Applications**, Marcel – Dekker Inc., New York, 1991, pg 349-377.

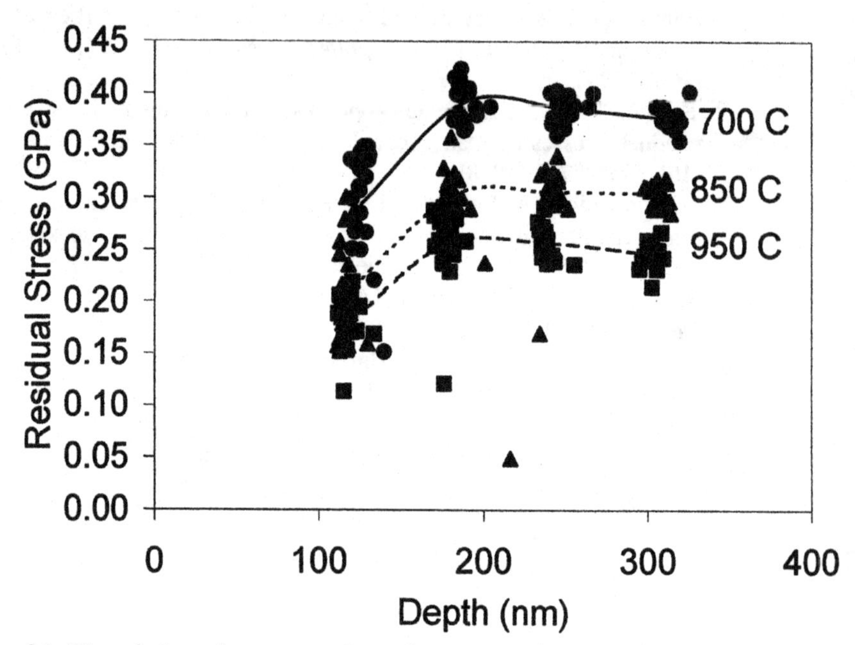

Figure 2A: Nano-indentation was performed on magnesium doped barium strontium titanate thin films. The results in (A) are for 10 mol% Mg doping in BST.

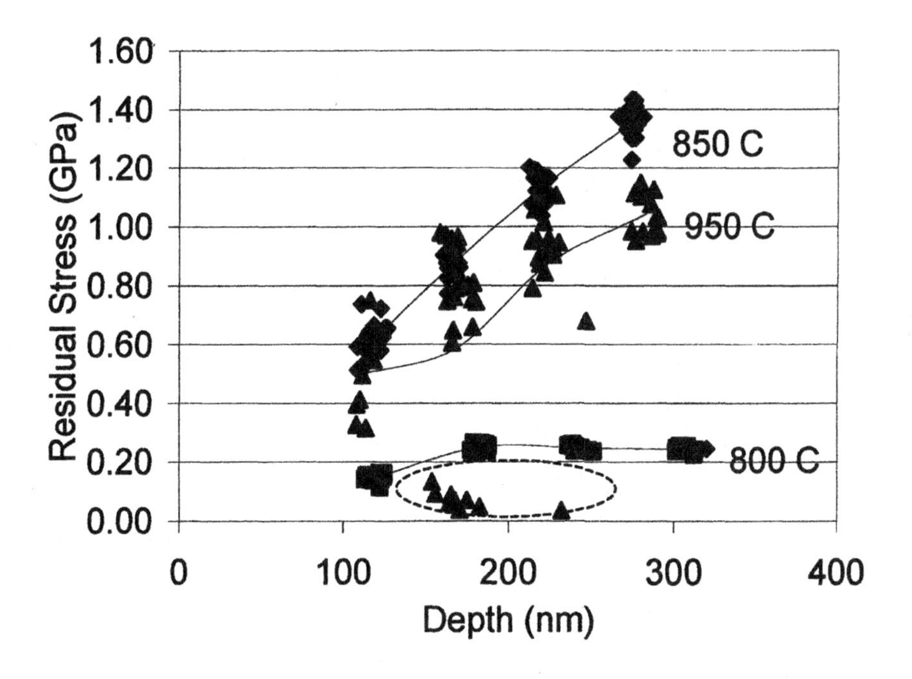

Figure 2B: Nano-indentation was performed on magnesium doped barium strontium titanate thin films. The results in (B) are for 5 mol% Mg in BST.

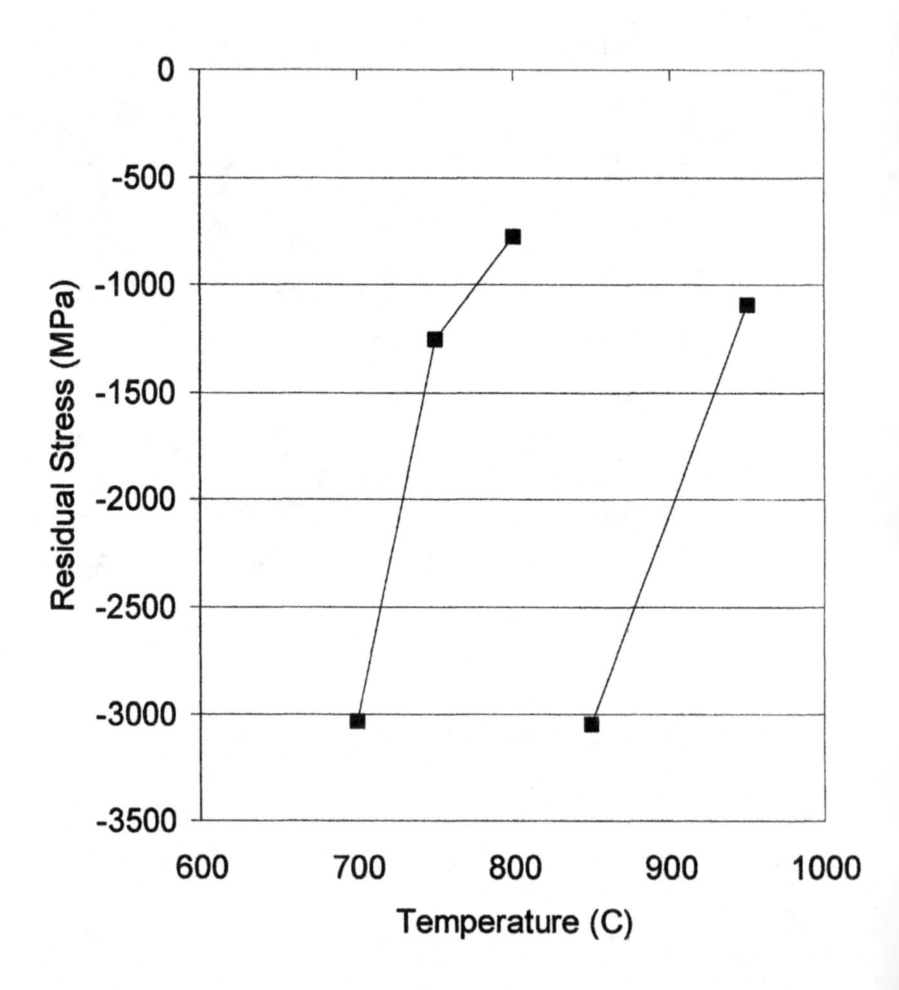

Figure 3A: RCA was used to measure the residual stress in magnesium doped barium strontium titanate thin films. The results in (A) are for 10 mol% Mg doped BST thin films.

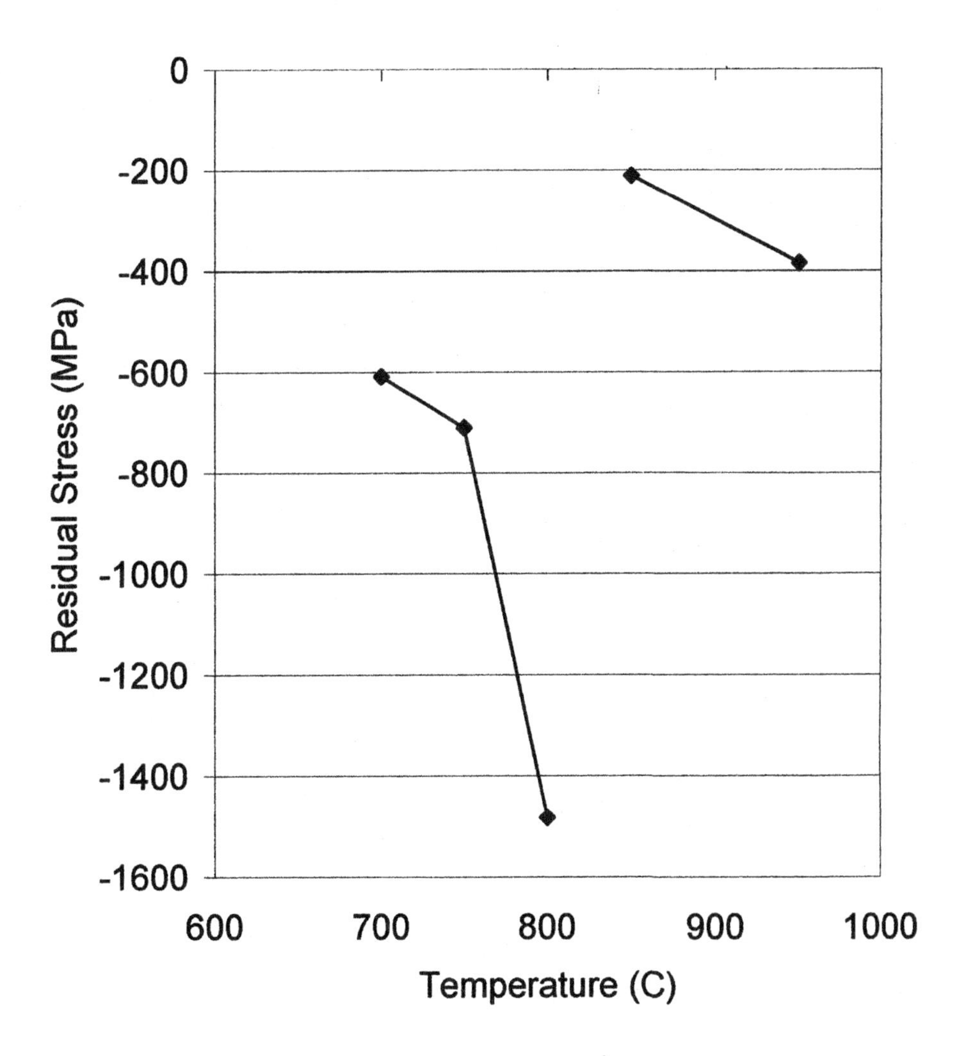

Figure 3B: RCA was used to measure the residual stress in magnesium doped barium strontium titanate thin films. The results in (B) are for 5 mol% Mg doped BST thin films.

STRENGTH DEGRADATION FROM CONTACT FATIGUE IN DENTAL CROWN COMPOSITES

Deuk Yong Lee
Daelim College of Technology
526-7, Bisan-Dong
Anyang 431-715, Korea

Se-Jong Lee
Kyungsung University
110-1, Daejeon-Dong
Busan 608-736, Korea

Il-Seok Park
Yonsei University
134, Shincheon-Dong
Seoul 120-749, Korea

Dae-Joon Kim
Sejong University
98, Gunja-Dong
Seoul 143-747, Korea

Bae-Yeon Kim
University of Incheon
177, Dohwa-Dong
Incheon 402-749, Korea

ABSTRACT

Hertzian cyclic fatigue properties of the glass-infiltrated alumina were evaluated in exact in vitro environment (artificial saliva) at contact loads (200-1000 N) to investigate indentation damage and strength degradation. At 200 N, no strength degradation was observed up to 10^6 contact cycles. As load increased from 200 N to 1000 N, the reduction in strength was found when the transition from ring to radial cracking occurred. The degree of strength degradation after critical cycling was more pronounced probably due to chemical reaction of the artificial saliva with the glass phase along the radial cracks introduced during large numbers of contact cycles.

INTRODUCTION

Alumina-glass composites prepared by melt infiltration have been considered as the method of choice for dental crown core materials because of their excellent biocompatibility, esthetics, wear resistance and chemical inertness as compared with traditional porcelain-fused-to-metal crowns.[1-4)] The use of melt-infiltration in all ceramic dental crowns provides near-net shape forming process (NNS) having low shrinkage for accurate fit, which is prerequisite for biomechanical components and dental crowns. NNS includes sintering of alumina at 1120°C to develop a skeleton of fused alumina particles and a subsequent infiltration of the porous structure with lanthanum aluminosilicate glass at 1100°C for the densification.[1-4)]

Several researchers[5-8)] reported that molar crowns should maintain their mechanical properties for extended period of time (>10^7 cycles) at cyclic contact loads (above 200 N) between opposing cusps of radii 2-4 mm in aqueous solutions because the lifetimes of dental restorations are limited by the accumulation of contact damage during oral function. Jung et al.[5] examined Hertzian cyclic contact fatigue properties of several dental ceramics using tungsten carbide ball in water and found that multi-cyclic damage accumulation limits the potential long-lifetime performance of dental ceramics. At large numbers of contact cycles the materials show an abrupt transition in damage mode, consisting of strongly enhanced damage inside the contact area and attendant initiation of radial cracks outside, resulting in the end of lifetime of the materials caused by rapid degradation in strength. Recently, Lawn[8)] analyzed ceramic-based multilayer structures using Hertzian cyclic loading for biomechanical applications to facilitate development of explicit fracture mechanics relations for predicting critical loads to produce lifetime-threatening damage. For dental crowns consisting of trilayers (veneer/core/dentin), the core is more susceptible to radial cracking rather than the weaker veneer because of stress concentration.

Although all-ceramic crowns have been investigated systematically by several researchers, most of Hertzian fatigue tests have been performed in water environment.[5-7)] In the present study, alumina-glass core materials were prepared and fatigue-tested using tungsten carbide ball (radius 3.18 mm) in exact in vitro environments (artificial saliva) at contact loads (200-1000 N) over millions of cycles to elucidate indentation damage and associated strength degradation.

EXPERIMENTAL PROCEDURE

Alumina powder (AL-M43, 99.9%, Sumitomo Chemical Co., Japan) was die-pressed into disks having a dimension of 20 mm in diameter and 2 mm in thickness, respectively and then isostatically pressed at 140 MPa. The green compacts were sintered for 2 h at 1120°C with a heating rate of 10°Cmin^{-1} and then furnace cooled. The La_2O_3-Al_2O_3-SiO_2 glass[2-4,9,10)] was prepared by melting the desired oxides in a platinum crucible at 1400°C, quenching the melt in water, and then grinding into a powder using a disc mill (Pulverisett 13, Fritsch GmbH, Germany). The glass powder-water slurry was placed on the partially sintered alumina preforms and infiltrated for up to 4 h at 1100°C with a rate of 30°Cmin^{-1} and then furnace cooled. The disk-type composites were polished to a 1 μm finish and then annealed for 10 min at 960°C to remove the possible residual stress. The final thickness of the composites was 1.7 mm.

Indentation tests were carried out using a WC sphere of radius of 3.18 mm up to loads in the range of 200 N to 1000 N at a preload of 19.6 N in artificial saliva environment. Artificial saliva having a pH of 5.2 was composed of KCl, NaCl, NaH_2PO_4, Na_2S and urea.[11)] The indenter is placed on the underside of the crosshead on a fatigue-testing machine (Instron 8871, USA) and the specimen centrally aligned along the load axis. The WC ball is lowered until light contact is made with the specimen. Cyclic tests were performed at a frequency of 10 Hz in haversinusoidal wave form at room temperature. The load was cycled between a specified maximum and small but nonzero minimum (19.6 N). All tests were done with the contact area immersed in artificial saliva. A minimum of five specimens was indented at each test.

Damage sites of gold-coated specimens were observed in reflection optical microscopy (ME-600L, Nikon Instech Co., Ltd., Japan) using Nomarski interference contrast. Indented specimens were then fractured using a flat-on-three-ball biaxial flexure testing (Instron 4202, USA). The specimens were broken using a biaxial strength fixture at a stress rate of 23 MPas^{-1}.[12)] The fractured specimens were also observed by Nomarski optical microscopy to investigate the fracture origin and path.

RESULTS AND DISCUSSION

Results of strength degradation as a function of numbers of fatigue cycles for the glass-infiltrated alumina are shown in Fig. 1. Open boxes at the left axes indicate standard deviation bounds of inert strengths for unindented specimens. Open symbols indicates failures originating from natural flaws and solid symbols implies failure traces coming from the contact damage. All specimens show an abrupt transition in strength with increasing number of cycles at various contact loads. Jung et al.[5] examined various material systems and reported that the glass-infiltrated alumina showed an intermediate behavior between brittle fracture in porcelain and glass and quasi-plastic response in yttria-stabilized zirconia polycrystals (Y-TZPs). Although the experiments were performed in artificial saliva environment, our results were in good agreement with those of ref. 5 conducted in water condition. However, strength degradation for the specimen tested under artificial saliva was more detrimental especially at higher contact load of 1000 N probably due to the chemical reaction with glass through cracks introduced during cyclic contacts.[5-8,11] Therefore, chemical attack of artificial saliva may be more severe than that of water.

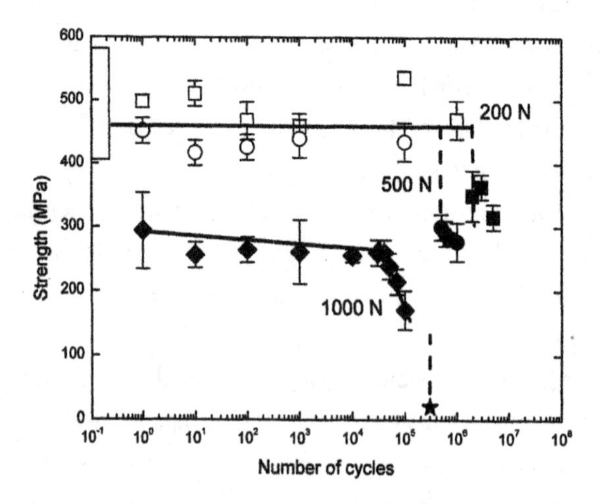

Fig. 1. Strength variation as a function of Hertzian fatigue cycles for glass-infiltrated alumina under different contact loads in artificial saliva.

At 200 N and up to 10^6 cycles, no cracks and impressions was observed for the composites, however, strength degradation occurred due to the detectable surface impression (P_Y) when the cyclic contacts were above $2x10^6$ cycles. The fracture path was observed at the contact edge, suggesting failure from a weakened interface at the boundary between the deformed plastic zone and its elastic surrounds.[5,6)]

At load of 500 N, P_Y and cracking (P_C) were found at cycles of 5×10^3 and 5×10^5, respectively. Ring crack of the composites (Fig. 2(a)) was changed to radial crack (Fig. 2(b)) associated with advanced quasi-plasticity when the cyclic contacts were above critical cycle of 10^6, as shown in Fig. 2, implying that more deleterious crack system introduced after large numbers of cycles leads to accelerated failure. Failure in Fig. 2(b) occurred at the contact interior (Fig. 3(b)) because the inner damage provided a sufficient intensity to initiate outwardly spreading radial cracks due to the accumulation of near contact damage generated by the penetrating indenter. The strength decrement above critical cycle was more severe than that reported earlier[5,6)] due to the existence of corrosive artificial saliva.

At load of 1000 N, the first strength drop was observed during the first contact, whose strength value was lower than that tested in water. P_Y and P_C were noticed at cycles of 3×10^3 and 4×10^4, respectively. The critical transition from ring crack to radial crack (Fig. 2(c)) was seen above 7×10^4 contacts, where the second strength degradation was examined (Fig. 1). At the cycle of 3×10^5, the specimen failed during the cyclic contact process.

Although the glass-infiltrated alumina is weaker than Y-TZP, it is also included in quasi-plastic ceramics as compared to the brittle materials such as porcelain and glass ceramics.[5-7)] Jung et al.[5)] argued that the key to the lifetime of the tougher dental ceramics such as the glass-infiltrated alumina and Y-TZP lay in the evolution of the quasi-plastic damage mode. The tougher glass-alumina composites suffered localized weakness in shear, weak interface between alumina and bonding glass, produced by the action of highly concentrated shear stresses. The interface sliding may initiate microcracks and then form macroscopic radial cracks after large numbers of cyclic contacts, resulting in drastic reduction in lifetime and strength. Therefore, contact damage introduced by stress concentration after enough contacts by a WC ball may be responsible for radial cracks and other subsurface embryonic cracks. The Hertzian contact damage may be more deleterious when the materials were exposed to artificial saliva rather than water.

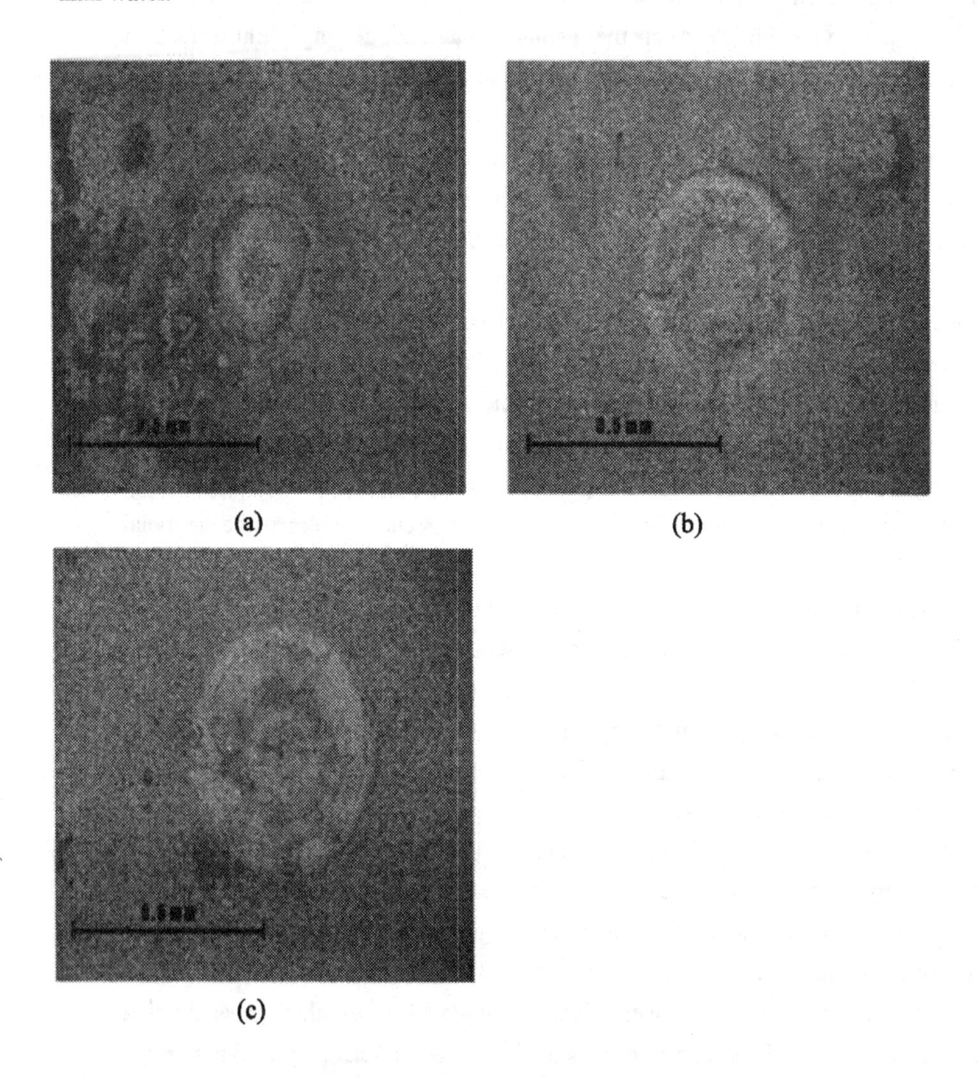

Fig. 2. Nomarski optical micrographs of Hertzian indentation sites of glass-infiltrated alumina after (a) 6×10^5, (b) 10^6 cycles at 500 N and (c) 7×10^4 at 1000 N, respectively.

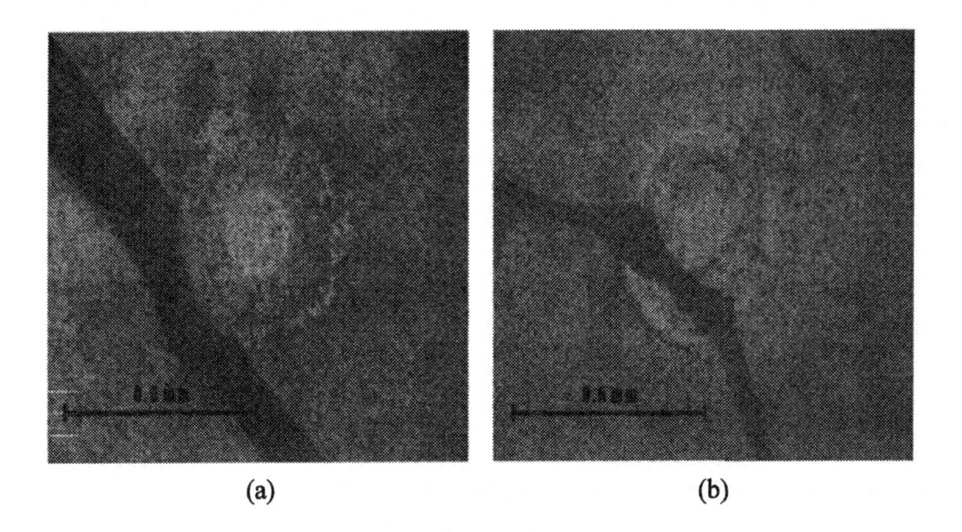

Fig. 3. Nomarski optical micrographs of Hertzian indentation failure sites of glass-infiltrated alumina after (a) 10^5 and (b) 10^6 cycles at 500 N, respectively.

Clinical variables such as masticatory force and cuspal curvature are closely related to Hertzian variables such as contact load and indenter radius.[5-8] Dental ceramic crowns should withstand masticatory forces of 200 N and more than 10^6 cycles at contacts between opposing cusps of radii of 2 to 4 mm.[6] Experimentally, one WC indentor (r=3.18 mm) and the specimen having a flat surface were used during the test instead of two opposing cusps.(r=2~4 mm) Therefore, contact damage inside the contact area may be enhanced due to the absence of curvature and stress concentration as a result of the use of flat specimen. The glass-infiltrated alumina observing no cracks and impression during cycling at 200 N may be highly applicable for the dental core materials under typical oral conditions because the composites were withstood even under severe test conditions.

CONCLUSIONS

The alumina-glass dental ceramics were cyclic fatigued in exact in vitro environment at contact loads (200-1000 N) to investigate indentation damage and strength degradation. The inert strength was maintained at 200 N and up to 10^6

cycles in test no cracks or impressions were observed, implying that the glass-infiltrated alumina may be applicable for the dental core materials under typical oral function. As load and numbers of contact cycles rose, the damage mode was changed from brittle fracture (ring crack) to deformation (radial crack) mode. Therefore, the brittle to plastic transition in the glass-infiltrated alumina was accelerated primarily by loads. Also, the extent of strength degradation after critical cycle was enhanced by the presence of artificial saliva probably due to the chemical reaction of artificial saliva with glass phase through radial cracks introduced during cyclic contacts.

ACKNOWLEDGEMENTS

This work was supported by grant No. R05-2000-000-00242-0 from the Basic Research Program of the Korea Science & Engineering Foundation.

REFERENCES

[1]W.D. Wolf, L.F. Francis, C-P. Lin and W.H. Douglas, "Melt-infiltration Processing and Fracture Toughness of Alumina-glass Dental Composites," *Journal of the American Ceramic Society*, **76** 2691-94 (1993).

[2]D-J. Kim, M-H. Lee, D.Y. Lee and J-S. Han, "A Comparison of Mechanical Properties of All-ceramic Alumina Dental Crowns Prepared from Aqueous- and Non-aqueous-based Tape Casting," *Journal of Biomedical Materials Research*, **53** 314-19 (2000).

[3]D.Y. Lee, D-J. Kim, B-Y. Kim and Y-S. Song, "Effect of Alumina Particle Size and Distribution on Infiltration Rate and Fracture Toughness of Alumina-glass Composites Prepared by Melt-infiltration," *Materials Science & Engineering A*, **341** 98-105 (2003).

[4]D.Y. Lee, J-W. Jang, M-H. Lee, J-K. Lee, D-J. Kim and I-S. Park, "Glass-alumina Composites Prepared by Melt-infiltration: II. Kinetic Studies," *Journal of Korean Ceramic Society*, **39** 145-52 (2002).

[5]Y-G. Jung, I.M. Peterson, D.K. Kim and B.R. Lawn, "Lifetime-limiting Strength Degradation from Contact Fatigue in Dental Ceramics," *Journal of Dental Research*, **79** 722-31 (2000).

[6]I.M. Peterson, A. Pajares, B.R. Lawn, V.P. Thompson and E.D. Rekow, "Mechanical Characterization of Dental Ceramics by Hertzian Contacts," *Journal*

of Dental Research, **77** 589-602 (1998).

[7]H. Cai, M.A.S. Kalceff, B.M. Hooks and B.R. Lawn, "Cyclic Fatigue of a Mica-containing Glass-ceramic at Hertzian Contacts," *Journal of Materials Research*, **9** 2654-61 (1994).

[8]B.R. Lawn, "Ceramic-based Layer Structures for Biomechanical Application," *Current Opinion in Solid State & Materials Science*, **6** 229-35 (2002).

[9]D.Y. Lee, J-W. Jang, H-K. Kim and B-Y. Kim, "Influence of Alkali and Alkaline Earth Addition on Thermal Expansion Coefficient of La_2O_3-Al_2O_3-SiO_2 Glass for Dental Ceramic Crowns," *Dental Materials*, submitted for publication.

[10]D.Y. Lee, D-J. Kim and Y-S. Song, "Properties of Glass-spinel Composites Prepared by Melt Iinfiltration," *Journal of Materials Science Letters*, **21** 1223-26 (2003).

[11]S-H. Lee, D-S. Ham, H-K. Kim, J-W. Jang and M-H. Kim, "The Effect of Burn-out Temperature and Cooling Rate on the Microstructure and Corrosion Behavior of Dental Casting Gold Alloy," *Journal of Korean Academy of Dental Technology*, **22** 73-82 (2000).

[12]Standard Test method for Biaxial Flexure Strength (Modulus of Rupture) of Ceramic Substrates, ASTM Designation F394-78, Annual Book of ASTM Standards Vol. 15.02, Section 15, 446-450. American Society for Testing and Materials, Philadelphia, PA, 1996.

APPLICATION OF FOCUSED ION BEAM MILLER IN INDENTATION FRACTURE CHARACTERIZATION

R. J. Moon, Z-H Xie, M. Hoffman and P. R. Munroe
School of Materials Science and Engineering
University of New South Wales
Sydney, NSW, 2052 Australia

Y-B Cheng
School of Physics and Materials Engineering
Monash University
Melbourne, VIC, 3800 Australia

ABSTRACT

A focused ion beam (FIB) miller was used to analyze the subsurface cracking behavior of two different α-sialon microstructures (equiaxed and rod-like) after Vickers indentation. The effect of grain morphology and grain orientation on the subsurface cracking behavior was observed and discussed.

INTRODUCTION

Vickers indentation testing is a simple method that has been widely used to measure mechanical properties of materials. Vickers indentation-induced deformation and cracking in brittle materials have been extensively investigated [1-5], where optically transparent materials (glasses, and single crystals) have been used to study the surface and subsurface evolution of different crack systems (i.e. median/radial and lateral) during the loading and unloading indentation cycle. To investigate the subsurface cracking behavior of optically opaque materials, previous investigations have used either ceramographic polishing [5] or the bonded-interface technique [4] for the preparation of subsurface cross-sections. It is noted that polishing may introduce further damage, while the presence of a bonded interface may affect the subsurface stress profile and thus the cracking behavior. The focused ion beam (FIB) miller has been used, in recent years, for analyzing subsurface fracture behavior of ceramics [6-8], as it enables subsurface cross-sections to be rapidly prepared through the area of interest with minimal damage. Combined with the compositional contrast arising from FIB imaging, the interaction between the subsurface microstructure and the different crack systems can be revealed.

In this present study, two α-sialon ceramics of differing grain morphologies (equiaxed and rod-like) underwent Vickers indentation tests, supported by the FIB miller, to explore the influence of grain shape and grain orientation on the indentation-induced subsurface cracking behavior.

EXPERIMENTAL

Equiaxed (EQ) and rod-like (RL) grained α-sialon microstructure samples were fabricated with the same chemical composition defined by the formula $Ca_xSi_{12-(m+n)}Al_{m+n}O_nN_{16-n}$, where x=m/2, m=2.6 and n=1.3, based on different sintering schedules. Powders of α-Si_3N_4, AlN, and $CaCO_3$ were mixed, cold-isostatically pressed into 25 mm pellets, then hot pressed at 1550 to 1750°C at 25MPa. Further details of the sample processing are described in an earlier work [9]. The two distinct microstructures are shown in figure 1, and microstructural features are listed in Table I. Hot-pressing resulted in a preferential grain orientation in the two α-sialon samples, with the long c-axis orientating perpendicular to the hot-pressing direction as confirmed by x-ray diffraction [9]. The texture within RL was stronger than EQ, possibly, as a result of grain growth.

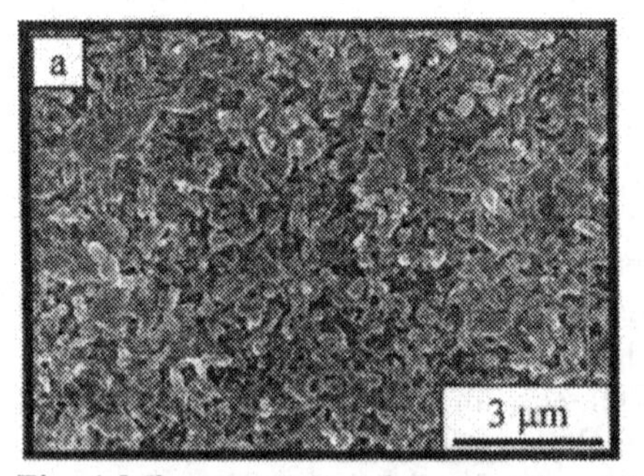

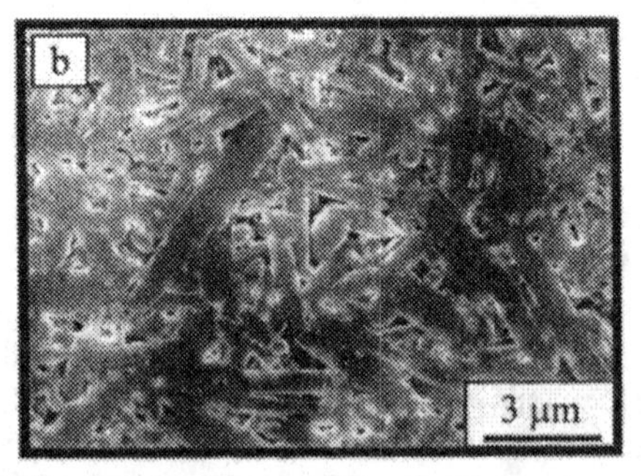

Fig. 1 Microstructure of Ca α-sialon (a) EQ and (b) RL materials. Surfaces were etched in molten NaOH and imaged by field emission scanning electron microscopy (from Xie et al [9]).

Table I. Microstructure features of α-sialon

Sample Identification	Average Grain Diameter [μm]	Aspect Ratio	Density [g/cm³]
EQ	0.35	1.1	3.19
RL	0.70	7.2	3.21

Samples were polished down to 1 μm diamond paste, as described in Ref [10]. All indentation tests were performed at ~25°C in laboratory air with a relative humidity of ~60%. Vickers indentation was conducted at loads of 300, 500 and 1000 g (Microhardness Tester, Model M-400-H1, Akashi Co., Japan). Vickers indentations were completed on the surface normal to and also parallel to the hot-pressing direction.

The FIB miller (FEI xP200, FEI Company, Hillsboro, OR 97124 USA) was used in two fashions. Firstly, the beam of Ga^+ cations, at a beam current of 70pA, was rastered over the sample surface providing a secondary electron image similar to that provided by scanning electron microscopes. The FIB image also gives compositional information, which is based on the secondary electron yield from a given phase due to the bombardment of these Ga^+ cations. FIB imaging was used to show the correlation between the microstructure and the crack profile. Secondly, cross-sections were prepared through the indentation-induced cracks via the FIB milling processes shown in figure 2, allowing for subsurface cracking behavior to be observed. The configuration of the milled region with respect to the indent used in this study is illustrated in figure 2.

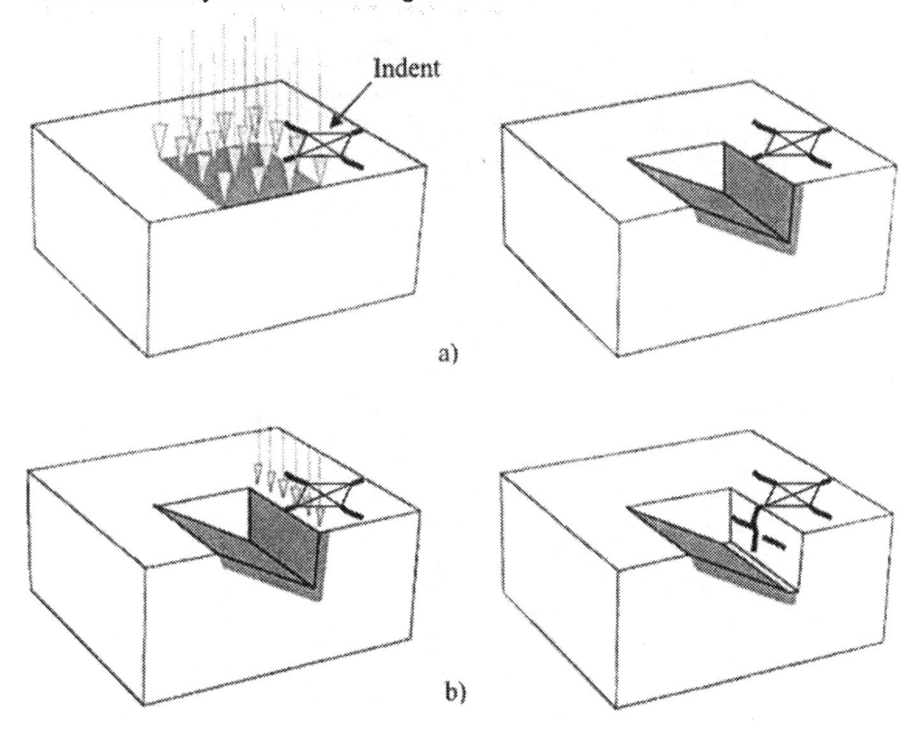

Fig. 2 Schematic of FIB milling technique: a) coarse milling at 6600pA beam current for "fast" material removal and b) fine milling at 1000pA beam current "cleans up" the final viewing surface. (Adapted from S. Chaiwan et al [11])

RESULTS AND DISCUSSIONS

Initial sub-surface observation

The initial sub-surface damage of both α-sialons microstructures investigated in this study was evaluated by conducting FIB analysis of polished surfaces. For both the EQ and RL microstructures, the amount of intrinsic subsurface damage was minimal as shown in Figure 3 for the RL microstructure. The lack of subsurface damage is believed to demonstrate the low amount of closed porosity in the material and that the damage introduced during the polishing procedures used [10] was minimal. The FIB imaging compositional contrast can be seen in figure 3, in which, on the cross-sectional view, the grey regions represent α-sialon and the light regions correspond to the glass phase. The dark region marked "a" is a large region of the glass phase which was preferentially milled during milling.

Fig. 3 FIB image showing a typical FIB milled region of the unindented RL sample showing the minimal amount of initial subsurface damage.

Indentation of equiaxed microstructure

Figure 4 shows the subsurface damage in the EQ microstructure that occurred just outside of the indent corners following Vickers indentation at loads of 300 g, 500 g and 1000 g. The results shown here are of indents on the surface normal to the hot-pressing direction, which were similar to those obtained for indents on the surface parallel to the hot-pressing direction. For the 300 g indentation load case (Fig. 4(a)), the radial crack is not clearly seen in the subsurface section, this may be due to the cross-section being cut too close to the indent corner or because the radial crack was highly localized at the surface. A number of microcracks parallel to the surface can also be seen. The lengths of the microcracks varied from less than the average grain length (0.3 μm) to ~2 μm. As the indent load increased to 500 g, the radial crack became larger and tended to propagate further into the subsurface (Fig. 4(b)). The size of the microcracks increased, possibly by microcrack coalescence, forming much longer cracks which began to resemble lateral cracks. At an indentation load of 1000 g, it can be seen that the radial and lateral cracks are well defined, having a large crack opening displacement (Fig. 4(c)). The propagation direction of each crack system appears to be affected by the presence of the other. In the region adjacent to the radial crack, the lateral crack was drawn to the surface. However, away from the radial crack, the lateral crack tended to propagate away from the surface. Likewise the radial crack kinked in the direction of the lateral cracks.

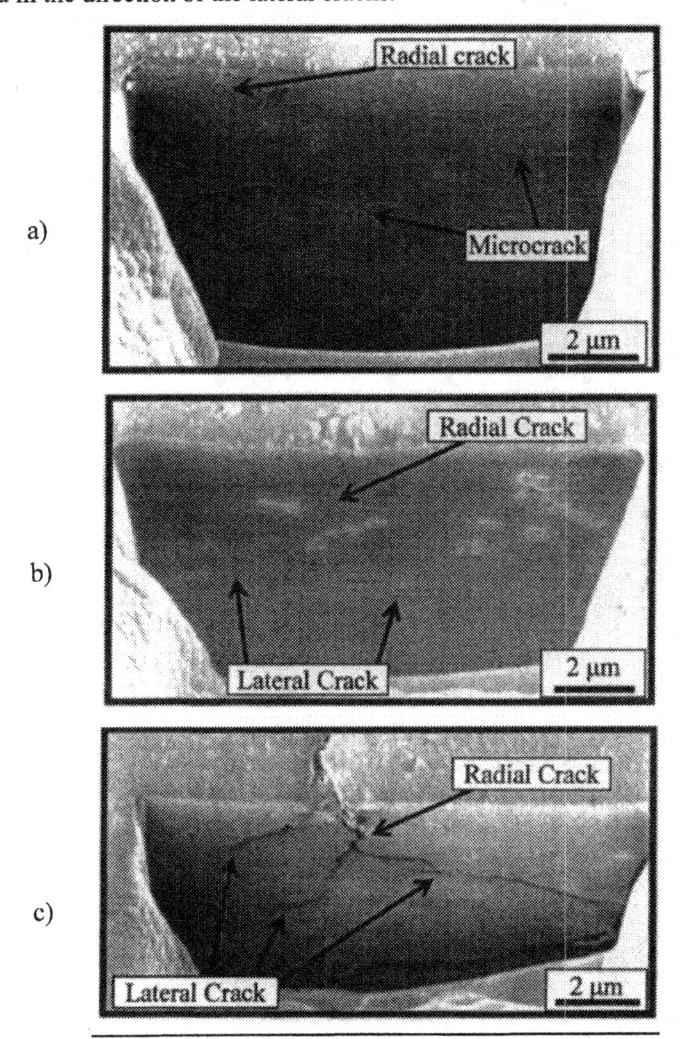

Fig. 4 FIB images showing subsurface damage of the EQ microstructure after Vickers indentation with loads of a) 300 g, b) 500 g, and c) 1000 g.

Indentation of rod-like microstructure

Figure 5 shows the subsurface damage in the RL microstructure that occurred just outside of the indent corners following Vickers indentation at loads of 300 g, 500 g and 1000 g. The results shown here are of indents on the surface normal to the hot-pressing direction. For the 300 g indentation load case (Fig. 5(a)), a well defined radial crack having a kinked fracture path and intragranular fracturing can be seen. Several microcracks appeared to have initiated from the radial crack. The microcracks formed along grain boundaries (i.e. through the glass phase), in which their lengths were comparable to the grain size. The path of the microcracking was approximately parallel to the surface, but was largely dependent on the orientation of the individual grains that these cracks extended along.

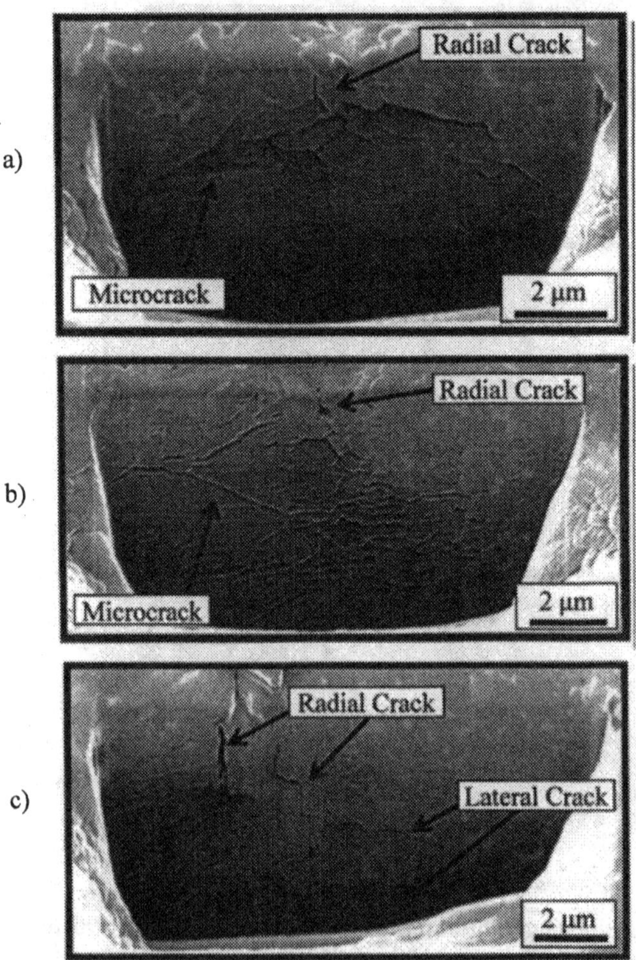

Fig. 5 FIB images showing subsurface damage of the RL microstructure after Vickers indentation, on the surface normal to the hot-pressing direction, with loads of a) 300 g, b) 500 g, and c) 1000g.

As the indentation load increased to 500 g, the formation of a radial crack can be seen and microcracks appear to have been initiated from the radial crack (Fig.

5(b)). The extent of grain boundary microcracking increased and in some cases it appeared that microcracking was not limited to the glass phase but also occurred in the α-sialon phase. For the 1000 g indent (Fig. 5(c)), the formation of radial cracks appeared to be more dominant than at the lower indentation loads. The radial crack path was torturous as the crack propagated around the grains. Lateral cracks, having a length of order of the elongated grains, appear to be created as a result of radial crack branching and microcrack propagation and coalescence along grain boundaries.

Orientation Influence

Subsurface analysis of Vickers indentations on the surface parallel to the hot-pressing direction was completed to investigate the effect of grain orientation on the subsurface cracks in the RL microstructure. Subsurface damage was examined following indentation with a 1000 g load. As shown in figure 6, the formation of a well defined radial crack dominates the subsurface fracture behavior. The formation of several microcracks can also be seen, however, the formation of well defined lateral cracks is missing.

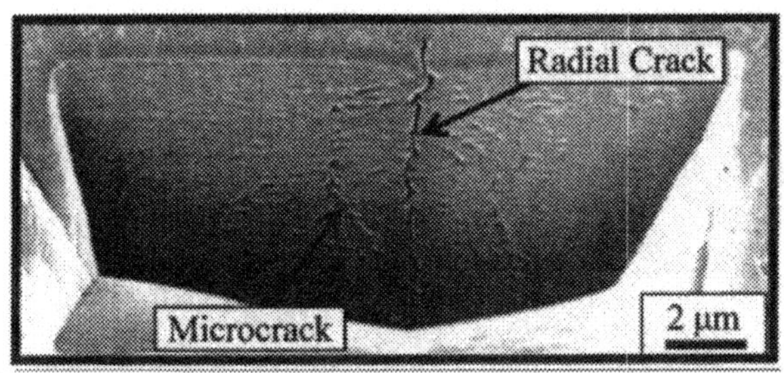

Fig. 6 FIB subsurface section of the 1000g load Vickers indent of the RL microstructure on the surface parallel to the hot-pressing direction.

The influence of the average grain orientation on the formation of the different crack systems can be observed by comparing figure 5(c) to figure 6. In figure 5 the average grain orientation is such that the long axis of the rod-like grains are horizontal, thus, one would expect an increased difficulty in forming radial cracks and a decreased difficulty in forming lateral cracks or microcracks that are oriented parallel to the surface. From figure 5(c) it is observed that the radial crack is kinked, suggesting difficulty in its formation, while lateral cracking (or microcracking parallel to the surface) can be seen to be less tortuous. In figure 6 the average grain orientation is such that the long axis of the rod-like grains is vertical, thus, one would expect an increased difficulty in forming lateral cracks or microcracks that are oriented parallel to the surface and a decreased difficulty in forming radial cracks. From figure 6 it is observed that the radial crack is straight and well defined, suggesting little difficulty in its formation, while lateral cracking (or microcracking parallel to the surface) can be seen to occur less often than in figure 5(c). Additionally, some microcracking also appeared to have formed parallel to the radial crack, which was not seen in figure 5(c), suggesting that microcracking in this orientation was facilitated by the average grain orientation. In general, the observed subsurface cracking behavior of these two orientations (Fig. 5(c) and Fig. 6) agreed with the idealized extreme cases, however, because the preferred orientation of the rod-like grains is not strong [9], some variability from the idealized case is to be expected.

Change of cracking behavior with location

The location of the milled region with respect to the Vickers indent was found to significantly influence the subsurface crack behavior. Figure 7 shows the location of two cross-section sites, one beneath the indent center and the other at the indent corner, following a 1000 g indentation on the RL microstructure. Underneath the indent center, no radial or lateral cracks were observed, however, fine microcracks parallel to the sample surface were observed (Fig. 7(b)).

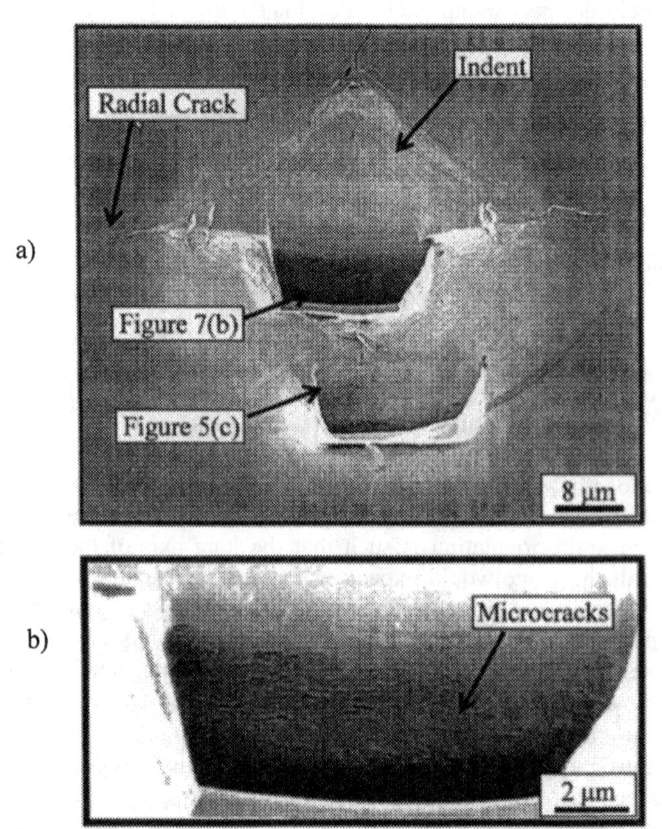

Fig. 7 FIB images of the 1000g load Vickers indent, on the surface normal to the hot-pressing direction, of the RL microstructure, showing a) the location of the two FIB cross-sections and b) cross-section under indent.

Outside the indent corner radial cracks and lateral cracks or microcracking of a size comparable to the length of elongated grains can be seen (Fig. 5(c)). It is believed that this "transition" between the two extreme cases occurred over a short distance (~4 μm) near the indent corner for this particular indent configuration (i.e., the material being indented, indenter shape, indenter load, radial crack shape, etc). It is also likely that the observed subsurface interaction between the radial and lateral cracks and the microcracking will vary as a function of the cross-section location along the length of the radial crack [5].

CONCLUSIONS

Vickers indentation tests combined with FIB analysis were used to evaluate subsurface contact-induced crack behavior. The effects of gain morphology and orientation on the subsurface crack growth in α-sialon ceramics were identified. The following conclusions can be drawn:

1) FIB milling can produce cross-sections through indentation induced cracks so that the subsurface cracking behavior can be observed.
2) The compositional contrast in FIB imaging allowed the interaction between microstructure and cracks to be revealed.
3) The grain morphology was observed to modify the subsurface cracking behavior induced by Vickers indentation. Intragranular fracture was dominant in both morphologies, however, for the rod-like grain morphology, the crack paths of radial and lateral cracks and the microcracking were observed to be influenced by the individual α-sialon grains. In contrast, the crack paths for the equiaxed morphology were not directly related to the grain structure. In general, the formation of lateral cracks occurred more easily in the fine equiaxed-grained α-sialon.
4) In the rod-like grained α-sialon, grain orientation significantly affected the radial and lateral crack and microcrack formation induced by Vickers indentation. For the situation in which the c-axis of the rod-like grains were oriented parallel to the indentation direction, a lower resistance to radial crack formation, but a greater resistance to lateral cracking and microcracking was observed.
5) The sub-surface cracking behavior induced by Vickers indentation was shown to vary as a function of location. The sub-surface cracking behavior directly beneath the indent consisted of microcracks, whereas outside the indent region, near one of the indent corners, radial and lateral cracks as well as microcracking were observed.

ACKNOWLEGDMENTS

This work was supported by an Australian Research Council Large Grant entitled "High Temperature Wear in α-Sialon Ceramics". α-sialon samples were prepared with the assistance of Dr. Yu Zhang at School of Physics and Materials Engineering, Monash University, Melbourne, Australia.

REFERENCES

[1]B. R. Lawn and R. Wilshaw, "Review of Indentation Fracture: Principles and Applications," *J. Mater. Sci.*, **10**, 1049-81 (1975).

[2]D. B. Marshall, B. R. Lawn and A. G. Evans, "Elastic/Plastic Indentation Damage in Ceramics: The Lateral Crack System," *J. Am. Ceram. Soc.*, **65** [11], 561-66 (1982).

[3]R. F. Cook and G. M. Pharr, "Direct Observation and Analysis of Indentation Cracking in Glasses and Ceramics," *J. Am. Ceram. Soc.*, **73** [4], 787-817 (1990).

[4]H. H. K. Xu and S. Jahanmir, "Simple Technique for Observing Subsurface Damage in Machining of Ceramics", *J. Am. Ceram. Soc.*, **77** [5], 1388-90 (1994).

[5]T. Lube, "Indentation Crack Profiles in Silicon Nitride," *J. Euro. Ceram. Soc.*, **21**, 211-18 (2001).

[6]N. Rowlands and P. Munroe, "FIB for the Evaluation of No-Semiconductor Materials" in *Proceedings of the 31st Annual Technical Meeting of the International Metallographic Society*, (eds. D. O. Northwood. E. Abramovici. M. T. Shehata and J. Wylie), pp.233-41, ASM International, Materials Park, Ohio, USA (1998).

[7]Z-H. Xie, M. Hoffman, R. J. Moon, P. Munroe and Y.-B. Cheng, "Scratch Damage in Ceramics: Role of Microstructure," *J. Am. Ceram. Soc.*, **86** [1], 141-48 (2003).

[8]Z-H. Xie, M. Hoffman, R. J. Moon, P. Munroe and Y.-B. Cheng, "Subsurface Indentation Damage and Mechanical Characterization of α-Sialon Ceramics," submitted to the journal of the American Ceramic Society.

[9]Z-H. Xie, M. Hoffman and Y. -B. Cheng, "Microstructural Tailoring and Characterization of a Calcium α-SiAlON Composition," *J. Am. Ceram. Soc.* **85**[4] 812-18 (2002).

[10]Z-H Xie, R. J. Moon, M. Hoffman, P. Munroe, and Y-B. Cheng, "Role of Microstructure in Grinding and Polishing of α-Sialon Ceramics," *J. Eur. Ceram. Soc.* **23** [13] 2351-60 (2003).

[11]S. Chaiwan, M. Hoffman, P. Munroe and U. Stiefel, "Investigation of Sub-Surface Damage During Sliding Wear of Alumina using Focused Ion-Beam Milling," *Wear* **252**, 531-39 (2002).

LOCAL T/M-PHASE AMOUNT DETERMINATION IN THE VICINITY OF INDENTATIONS AND SCRATCHES IN ZIRCONIA

Michael T. Dorn and Klaus G. Nickel

Eberhard-Karls-Universität Tübingen

Institute for Geosciences and Applied Mineralogy

D-72074 Tübingen, Germany

1 ABSTRACT

From Raman spectra next to Vickers indentations in 3Y-TZP there is strong evidence for grain orientation taking place during the indentation process. The grain orientations are related to the stress field produced by the indenter during loading process. The evidence is based on the variation of Raman band intensities with the direction of laser beam polarisation.

Each Raman band has a different degree of intensity variation with laser polarisation direction. Hence the calculation of phase amounts of tetragonal and monoclinic phase in zirconia, which uses band intensities changing with incident laser direction, will record a variation, which is caused solely by grain orientation.

In our presentation we show the persistence of the phenomena next to scratches and we discuss new equations for quantitative phase determination, which are more robust to such textural effects.

2 INTRODUCTION

The toughening mechanism of Zirconia based on the tetragonal-monoclinic ("t-m") phase transformation has been studied for many years [1] and was observed in a number of Zirconia modifications.

In the vicinity of hardness indentations at room temperature t-m transformations take place because of hydrostatic and deviatoric stresses at the surface and in the bulk material, which also cause micro/macro-cracking and plastic deformation. Next to indentations a strong uplift of the surface is observed, which is attributed primarily to the volume increase during the phase transformation and to a minor extend to the transport of material out of the contact area [2].

The determination of the extend of microstructural changes around indentations is of technological interest, because it will allow to evaluate the behavior of transformation toughened ceramics under non-uniform stresses, which are typical for wear or machining processes. Raman spectroscopy is very useful tool to determine the lateral extend of the transformed region and to estimate the amount of monoclinic phase.

However, the Raman band intensity and therefore the quantitative or semi-quantitative estimation of the amount of monoclinic phase in a partially transformed region [3-7] can be hampered by the band intensity dependence on the relative orientation of the crystal axis with respect to the polarization of the incident light and the polarization of the detected Raman light. This dependence has been demonstrated for the tetragonal [8] as well as for the monoclinic phase [9].

Polycrystalline samples have usually random orientated crystals and the Raman spectra show the sum of all possible excitations as long as the analyzed area is bigger then the grain size. The orientation effects will then cancel each other. If there is a true orientation of grains in a textured material, this will be no longer the case. It is known that residual stresses can cause a Raman signal peak shift, which in turn can be used to infer the level and distribution of residual stresses [10] [11]. There is thus a second reason for band intensity variations by band broadening, which needs to be considered. The present investigation explores mainly the vicinity of Vickers hardness indentations to elucidate the dependencies of Raman signals in stressed partially transformed Zirconia to allow a fast and non-destructive evaluation of phase and stress relations. First investigations the vicinity of scratches were also done.

3 EXPERIMENTAL PROCEDURE

Dense high purity 3Y-ZrO_2 samples from the Zentrum für Zahn-, Mund- und Kieferheilkunde of the Eberhard-Karls-Universität Tübingen with an average grain size of about 0.6 μm and a density >6.05 g/cm³ were used. They were polished to a 1 μm finish using diamond pastes. The indentations were made using an Instron 4502 universal test machine with a single load (50N) obtained after a constant loading (0.01mm/min) and fast unloading (1mm/min) speed using a Vickers indenter. Scratches on similar samples were produced with a load of 25N by a motorized x-y-z stage using an Knoop indenter.

A LabRam-2 micro-Raman spectrometer (Dilor, France) with excitation wavelength of 488nm (Ar^+ ion laser) was used in backscattering geometry to analyze the phase transformations to the monoclinic phase after indentation. To get comparable results, we used the same parameters for all investigations, including generated laser intensity (~100mW), objective magnification (50xULW) and focus on the sample. The lateral resolution of this objective is about 2 μm. There was no observable re-transformation of the transformed m-phase by sample heating with the applied laser intensities.

Monoclinic ZrO_2 (space group C^5_{2h}) has a spectrum with 18 Raman active modes [12]. The spectrum of tetragonal Zirconia (space group D^{15}_{4h}) has six typical bands of different Raman active modes ($A_{1g} + 2B_{1g} + 3E_g$) [10]. These bands have been assigned to different bonds [13] [14]. Because of the overlapping of several bands not all predicted monoclinic Raman modes can be discriminated. Those strongly overlapping bands are usually treated as one signal with two peaks.

Significant are the bands at 181/190 cm^{-1}, 379 cm^{-1} and 538/557 cm^{-1}. In samples with both t and m modifications we have overlap of the tetragonal bands at 322 and 464 cm^{-1} with monoclinic bands at 333 and 475 cm^{-1}. The imperfect superposition of these bands makes them visible as shoulders in bands or generate band broadening. Other monoclinic Raman bands of less intensity likewise influence the spectrum by modification of underground and band widths.

This is shown in Fig. 1. Here the spectrum of pure tetragonal ZrO_2 comes from an area far from the indentations and acts as a reference. At a point near an indentation, where monoclinic ZrO_2 is present in addition, intensity variations, peak shifts and peak broadening are observed. Subtracting the intensities of the mixed t-m-spectrum from the pure t-spectrum helps in the analysis, intensities below the y = 0 line indicate the shrinking of tetragonal band intensities, those above the appearance of monoclinic phase bands. In addition, representative for the determination of any band intensities, this of the tetragonal band at 265 cm^{-1} is shown. This intensity is defined as the area enclosed by the spectra and the baseline of two boundaries with the spectra. These boundaries are usually located at the zero values of the difference spectrum.

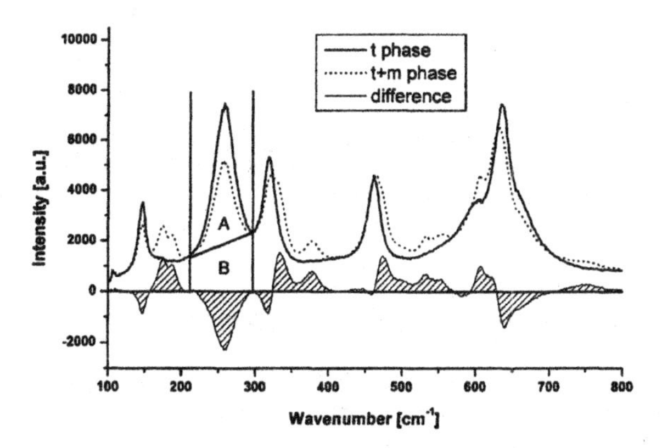

Fig. 1: Analyses of mixed t-m-Zirconia Raman spectrum by plotting the differences relative to the pure tetragonal phase.

However, the analyses as shown above may be biased by preferred orientation of grains. Therefore, experiments were carried out, in which the incident laser beam was rotated. Rotation of the linearly polarized laser beam was achieved by a retardation- (or phase-shifter-, half-wave-) plate. The thickness of such a plate is chosen to obtain a phase difference (retardation of the slow ray by comparison with the fast ray at emergence) of n times $\lambda/2$ (n = 1,2...) for the wavelength λ of the laser used and is placed directly behind the laser source. The incident linearly polarized beam emerges after its passage through the plate as a linearly polarized beam with a rotated polarization direction controlled by the position of the retarders principal plane.

We concentrate on linescans directly next to the indentation edge (A, length 80μm) and linescans perpendicular to the scratch edge with point spacings of 2 μm (Fig. 2). The starting (0°) orientation of the beam for both investigations is also shown in Fig. 2. Rotation was done at each spot of the linescans rotating anticlockwise in steps of 10° for the indentation test and 5° for the scratch test.

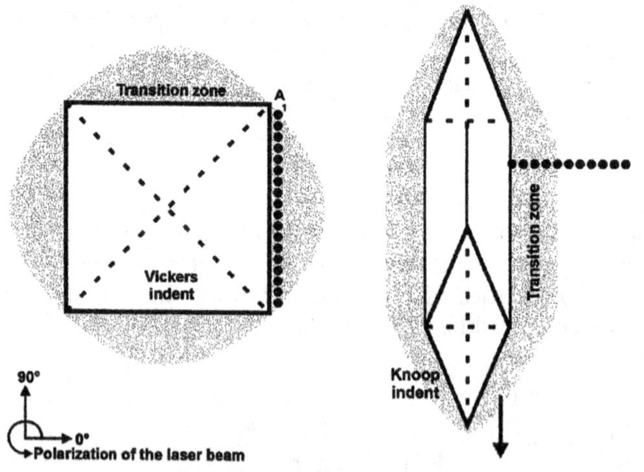

Fig. 2: Position of the linescan relative to indentation (left): A, starting at points marked "1" and relative to the scratch (right). The polarized laser beam was rotated at each point anticlockwise in steps of 10° for the indentation, 5° for the scratch test..

4 RESULTS

4.1 Raman band intensity changes next to indentations

In Fig. 3 we compare the intensity change of the Raman spectra at the starting point of linescan A (close to the indentation corner) at rotation angles of 30° and 110°. These particular angles are shown, because the tetragonal band at 265 cm^{-1}

had a maximum intensity when the external laser polarization was rotated 110° and a minimum at the rotation angle of 30°.

The intensity changes for the monoclinic bands at 190, 379, 380 and 480 cm^{-1} are not as strong but still significant as evidenced by the difference spectra (Fig. 3).

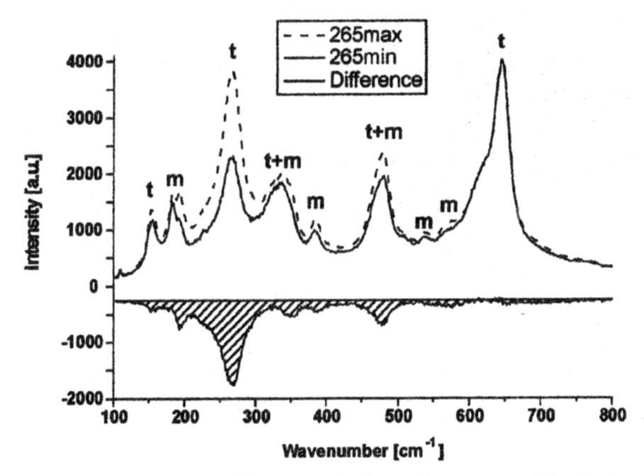

Fig. 3: Raman spectra and difference plot from the corner of the indentation (point "1" of linescan A in Fig. 2) at rotation angles of 30/110°.

In Fig. 3 it can be seen that the m-band intensity at 190 cm^{-1} changes with the tetragonal 265 cm^{-1} band, while the m-band intensity at 181 cm^{-1} does not. This feature of synchronous band intensity changes has been observed in all our rotational spectra. Also a near constant intensity of the tetragonal 648 cm^{-1} band can been found in Fig. 3. This band was therefore used to calculate the intensity ratios discussed below. We need intensity ratios to normalize and compare the intensities of spectra from different spots and samples.

Using the t-band with maximum change (265 cm^{-1}) normalized to the band with the lowest change (648 cm^{-1}) is thus the preferred result description. A low ratio $R = 265/648$ cm^{-1} indicates a high reduction of the tetragonal band and vice versa.

Now, it is also possible to display the reduction of this ratio in comparison to the pristine, unindented or unscratched surface of the samples samples.

The plot of the intensity ratio reduction at the linescan A (Fig. 2) has 40 points at a length of about 80 μm with an increment of 2 μm. At each point rotation about 10° for was performed and is shown in Fig. 4.

It is obvious from Fig. 4 that each point has a strong variation of the intensity ratio with the beam orientation.

However, the nominally equivalent points at the corners of the indentation (start and end point of the scan) do not have their minima and maxima at the same angle. The relation is symmetrical with low ratios (high tetragonal band reduction) for the starting point at 30° and -30° for the end point.

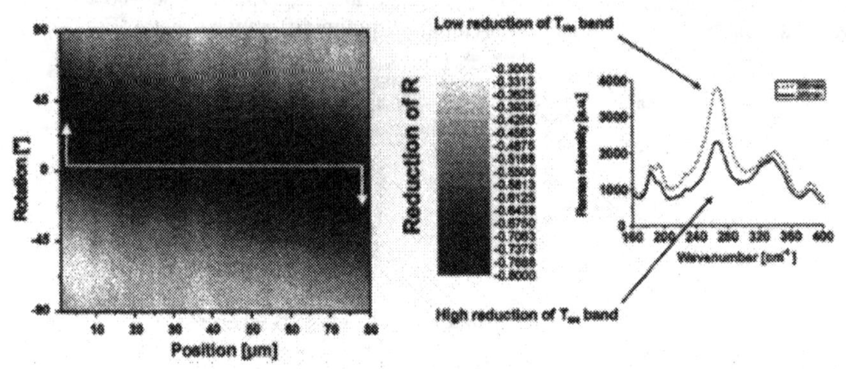

Fig. 4: Intensity ratio 265/648 cm^{-1} variation along the indentation edge (linescan A of Fig. 2) as a function of rotation angle of the polarized beam. The lowest values represent the highest reduction of the tetragonal band and therefore of the ratio R.

4.2 Raman band intensity changes next to scratches

A similar investigation was done in regions next to scratches (see Fig. 2). The results for two scratching loads are shown in Fig. 5.

Far away from the scratch, there is no change of R at any used loading and therefore of the intensity of the tetragonal Raman band at 265 cm^{-1} during the rotation of the incident laser polarization. Approaching the scratch, the ratio R shows lower values at 0° (a polarization orientation perpendicular to the scratch and the scratching direction). In a region of about 5μm directly next to the scratch, this minima rotates about 90° and appears at a direction of 90°, parallel to the scratch.

There is a strong dependence between the reduction of R and the scratch load. The average reduction directly at the scratch edge increases with increasing loads and the variation of intensities during the rotation is also increasing.

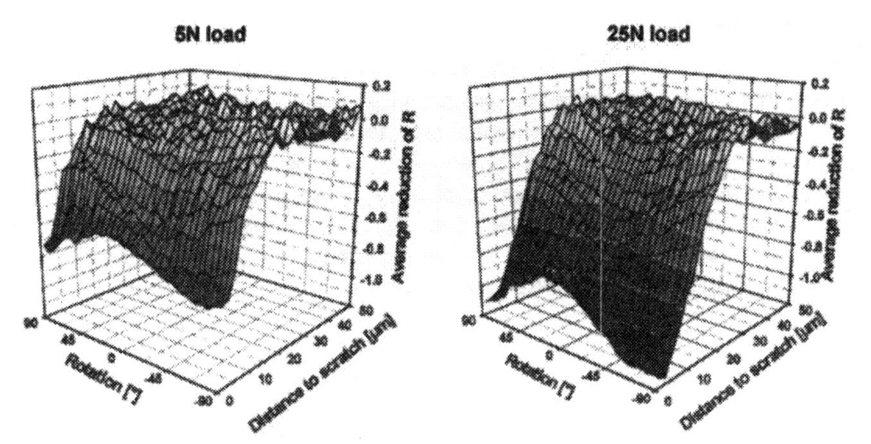

Fig. 5: Intensity ratio 265/648 cm^{-1} reduction (in comparison to unindented, unscratched samples) along the scratch edge (see Fig. 2) as a function of rotation angle of the polarized beam. The lowest values represent the highest change (reduction) of the tetragonal band.

5 DISCUSSION

5.1 Reasons for oriented band intensity reductions

A general dependence of Raman band intensities on crystal orientation is plain, because the very principal of this spectroscopy is the inelastic dispersion of light by the interaction with molecular or lattice vibrations. Anisotropic bodies like crystals have uneven atom densities in different directions and hence anisotropic polarisability and vibrational properties.

The dependence of the intensity of Raman bands and their ratios on the crystal orientation of ZrO_2 is known. For different scattering configurations, Merle et al. [8] associated and discussed the mode assignments for Raman bands of t-ZrO_2. In contrast to the x(zz)-x and the x(zy)-x polarization measurements, the A_{1g} mode at 265 cm^{-1} vanishes in z(xx)-z configuration and a slight rise of the underground up to 250 cm^{-1} was observed.

As shown above we observed a strong dependency of the incident laser polarization orientation despite the polycrystalline nature of the sample. We furthermore observed a slightly rising underground at lower wavenumbers in directions with low intensity of the tetragonal 265 cm^{-1} band, the increase was a function of the distance to the indentation. Thus the feature could be explained by a preferred orientation of grains within the vicinity of hardness indentations.

5.2 Stress field next to indentations

The stress field caused by the indentation process, which is responsible for non-elastic deformation of the region around Vickers impressions, has been modeled by Galanov [15] . In Fig. 6 the surface trajectories of the principal stresses σ_1 and σ_2 from this work are shown. The directions of such stresses correspond to the directions associated with the directions of the extreme values of R and therefore the intensities of the t-band at 265 cm^{-1}. The distribution of the stress level of contact stresses is quite different and has the highest values in the diagonal lines corresponding to the edges of the indenter [15].

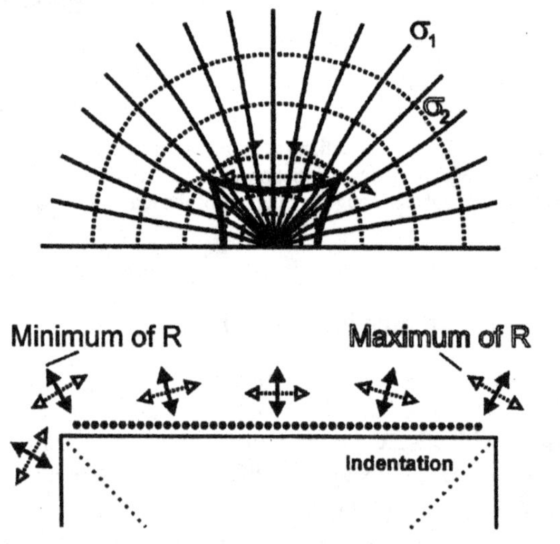

Fig. 6: Surface trajectories of σ_1 and σ_2 in comparison of the measured minimum and maximum values of R next to the indentation edge.

Our findings are therefore compatible with a model, in which the grains around Vickers indentations become at least partly aligned during plastic deformation.

5.3 Stress field next to scratches

From the attributions of σ_1 and σ_2 to the extreme values of R next to indentations, it is also possible to model the stress field next to scratches (Fig. 7) by the minimum and maximum values of R.

In contrast to the static indentation tests, where a change on the extreme values of R along the indentations edge was observed, only a specific orientation was detected in the scratch test.

For far distances to the scratch, the minimum of R and therefore the direction of σ_1 can be determined perpendicular to the scratch. The direction of σ_2 is already perpendicular to the direction of σ_1 and parallel to the scratch edge. A difference occurs directly at the edge of the scratch, where these directions are rotated by 90°.

Presently we have no explanation for this change in direction.

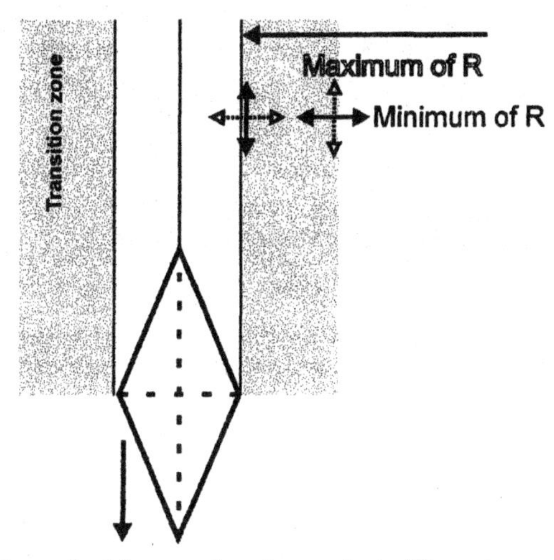

Fig. 7: Measured minimum and maximum values of R next to scratches.

5.4 Calculation of the monoclinic fraction

The following equation has been suggested to calculate the monoclinic phase content [3] [5]:

$$c_m = \frac{I_m^{181} + I_m^{190}}{F \bullet \left(I_t^{147} + I_t^{264}\right) + I_m^{181} + I_m^{190}} \qquad \{1\}$$

In eq. {1} I_m is the integral value of the monoclinic bands at 181/190 cm^{-1} and I_t of the tetragonal bands at 147 and 265 cm^{-1}. The factor F was used to correlate the Raman intensities to the XRD intensities and was 0.97 [3].

Massive changes of the t-band intensity at 265 cm^{-1} will obviously directly influence the monoclinic fraction calculated from eq. 1. In Fig. 8 we calculated these values for linescan A (directly next to the indentation edge).

The calculated monoclinic phase content differs between 15 and 50%. Close to the indentations corner (point 40) the maximum value is about 50%, the minimum about 35%. The difference is the direct consequence of the intensity change at a single position. In Fig. 6 we have shown that the angle for extreme band intensities varies along the indentation from ~30/-150° to -30/150°. Accordingly a change in the calculated phase amounts is following this pattern in Fig. 8.

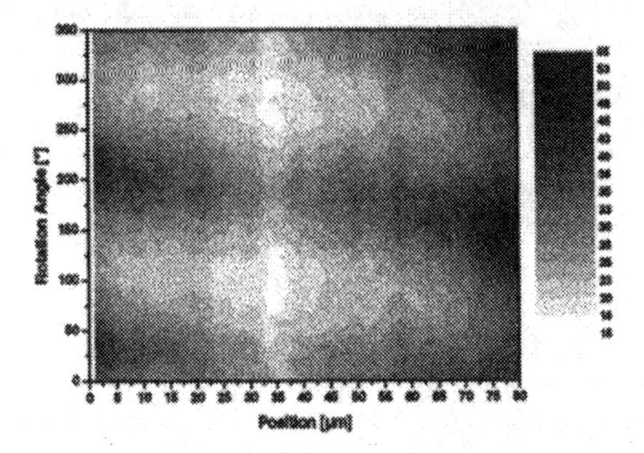

Fig. 8: Calculated monoclinic phase content [%] for scan A in dependency of the rotation angle of the incident laser plane relative to the indentations edge.

Clearly the differences render equation {1} as not applicable. An evaluation the monoclinic fraction must use a reference band for $t\text{-}ZrO_2$, which is not biased by the rotation of the laser beam polarization direction. As shown in Fig. 3 the 648 cm^{-1} band fulfils this condition. We recommend therefore to use equation {2} for the calculation of the amount of monoclinic Zirconia:

$$c_m = \frac{I_m^{181} + I_m^{190}}{(K * I_t^{648}) + I_m^{181} + I_m^{190}} \qquad \{2\}$$

To get similar values of the monoclinic phase, it is possible to customize eq. {2} by implementing a factor K. In our investigation it was about K=0.42. The strongly differing spread of the results (Fig. 9) is simply the expression of the nearly cancelled rotational dependence of the tetragonal band signal in eq. {2} and the low rotational dependency of the monoclinic bands.

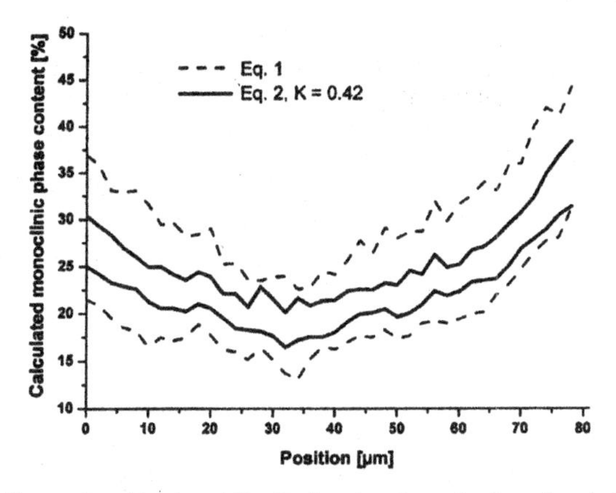

Fig. 9: Comparison Eq. 1 and Eq. 2, showing the reduction of variation in calculated monoclinic phase contents along linescan A of Fig. 2.

The similarity of the shapes is obvious. The contents calculated by eq. {2} correspond well to the results calculated by eq. {1}, using average tetragonal band intensities.

An accordingly calculated plot along linescan A is shown in Fig. 10.

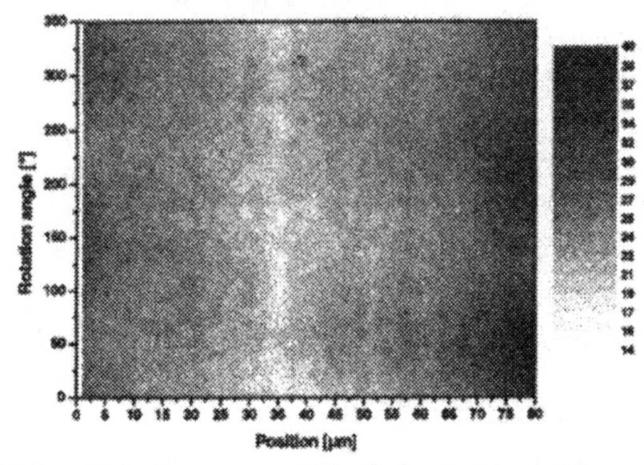

Fig. 10: Monoclinic phase content [%] calculated with eq. {2}.

In contrast to the calculation with the strong orientation dependent tetragonal Raman band at 265 cm^{-1} in Fig. 8, the monoclinic phase content is now seen as a clear variation with low values in the center of the scan to high values at the end points corresponding to the edges of the indentation.

It is therefore identical to use eq. {2} and eq. {1} with the mean value of rotational plots. The use of depolarized incident light achieved by depolarizers or scramblers or circular polarized light via quarter-wave-plates will also lead to the same result using either eq. {2} or {1}.

Nonetheless the measurement with polarized light has advantages: A difference of values calculated from eq.{1} and {2} from a randomly chosen spot points towards a orientation of grains. Thus it provides additional information on the microstructure of the material and eq. {2} avoids misinterpretation of phase contents in orientated samples.

6 CONCLUSION

We have investigated the determination of transformed monoclinic phase content after Vickers indentation on Y-TZP with Raman spectroscopy. Rotation of the angle of incident of the polarized laser beam caused a strong variation of the intensity of the tetragonal Raman band at 265 cm^{-1} at any spot close to the indentation. This is taken as evidence for an orientation of grains in the vicinity of hardness indentations.

The angles, at which maximum intensities of this band are detected, vary systematically along the indentation. The variation is related to the trajectories of stress lines calculated for indentation processes. The indentation thus seems to cause an orientation of grains according to the stress field.

At scratching tests, a strong dependence of the ratio R can be detected as well. The orientation of the extreme values change directly at the scratch edge, in contrast to areas more far away. Both effects become stronger at higher scratching loads.

The amount of transformed zirconia is not equal along the edges of hardness indentations. The monoclinic phase content is highest at the corners of the indentation. This is related to the maximum contact stresses calculated for the indentation.

The orientation dependent intensity of the tetragonal Raman band at 265 cm^{-1} leads to a large spread of calculated monoclinic phase contents for any single point using a widely applied formula [3].

There is a orientation depended intensity also for some monoclinic bands but it seems to be less pronounced.

As a more robust method of calculating monoclinic phase contents we recommend the use of eq. {2}, in which the normalization is done on the basis of the Raman band at 648 cm^{-1}, which is not orientation dependent.

Polarized Raman spectroscopy allows therefore the detection of phase contents in zirconia as well as of microstructural information.

[1]R.H.J. Hannink, P.M. Kelly and B.C. Muddle, "Transformation Toughening in Zirconia-Containing Ceramics", *J. Am. Ceram. Soc.* **83** [3] 461-87 (2000).

[2]A. Celli, "AFM, a tool for investigating indentation damage in ZrO2", *J. Am.Ceram. Soc.* **60** [8] 87-91 (1999).

[3]D.R. Clarke, "Measurement of the Crystallographically Transformed Zone Produced by Fracture in Ceramics Containing Tetragonal Zirconia", *J. Am. Ceram. Soc.* **65** [6] 284-8 (1982).

[4]G. Behrens, G.W. Dransmann and A.H. Heuer, "On the Isothermal Martensitic Transformation in 3Y-TZP", *J. Am. Ceram. Soc.* **76** [4] 1025-30 (1993).

[5]M.S. Kaliszewski, *et al.*, "Indentation Studies on Y2O3-Stabilized ZrO2: I, Development of Indentation-Induced Cracks", *J. Am. Ceram. Soc.* **77** [5] 1185-93 (1994).

[6]S.A.S. Asif, D.V.S. Muthu, A.K. Sood and S.K. Biswas, "Surface Damage of Yttria-Tetragonal Zirconia Polycrystals and Magnesia-Partially-Stabilized Zirconia in Single-Point Abrasion", *J. Am. Ceram. Soc.* **78** [12] 3357-362 (1995).

[7]V. Sergo, D. Clarke and W. Pompe, "Deformation Bands in Ceria-Stabilized Tetragonal Zirconia Alumina .1. Measurement of Internal Stresses", *J. Am. Ceram. Soc.* **78** [3] 633-40 (1995).

[8]T. Merle, R. Guinebretiere, A. Mirgorodsky and P. Quintard, "Polarized Raman spectra of tetragonal pure ZrO2 measured on epitaxial films", *Phys. Rev. B.* **65** [14] art. no.-144302 (2002).

[9]M. Ishigame and T. Sakurai, "Temperature Dependence of the Raman Spectra of ZrO2", *J. Am. Ceram. Soc.* **60** [7-8] 367-9 (1976).

[10]H. Arashi and M. Ishigame, "Raman spectroscopic studies of the polymorphism in ZrO_2 at high pressures", *Phys. Stat. Sol. (a)* **71** 313-21 (1982).

[11]J.A. Pardo, R.I. Merino, V.M. Orera, J.I. Pena, C. Gonzalez, J.Y. Pastor and J. Llorca, "Piezospectroscopic study of residual stresses in Al2O3-ZrO2 directionally solidified eutectics", *J. Am. Ceram. Soc.* **83** [11] 2745-52 (2000).

[12]X. Zhao and D. Vanderbilt, "Phonons and lattice dielectric properties of zirconia", *Phys. Rev. B* **65** 2002).

[13]P. Duran, F. Capel, C. Moure, A.R. Gonzalez-Elipe, A. Caballero and M.A. Banares, "Mixed (oxygen ion and n-type) conductivity and structural characterization of titania-doped stabilized tetragonal zirconia", *Journal of the Electrochemical Society* **146** [7] 2425-34 (1999).

[14]P. Li and I.-W. Chen, "X-ray-absorption studies of zirconia polymorphs. III. Static distortion and thermal distortion", *Phys. Rev. B* **48** [14] 10082-9 (1993).

[15]B.A. Galanov and O.N. Grigoriev, "Deformation and fracture of superhard materials under contact loading", *Problemy Prochnosty* [10] 36-42 (1986).

INVESTIGATION OF THE CREEP BEHAVIOUR OF A CERAMIC-BASED COMPOSITE, USING NANOINDENTION TESTING OF THE CREEPING CONSTITUENT

R. Goodall, A. Kahl and T. W. Clyne
Cambridge University Dept. of Materials Science & Metallurgy
New Museums Site, Pembroke Street
Cambridge, CB2 3QZ. UK

J A Fernie
TWI
Granta Park, Great Abington
Cambridge, CB1 6AL, UK

ABSTRACT

Barrikade® is a novel low density, fire-resistant material, consisting of particles of vermiculite held together by a water soluble sodium silicate binder. This material is being developed for use in a range of potential applications, including a core material in fire doors. The creep behaviour of the material, critical for many of these applications, has been investigated using a variety of techniques.

Creep of the composite is expected to be dominated by behaviour of the binder. Since this constituent can only be produced in small samples if it is to have a structure representative of that in the composite, a nanoindentation technique was used to investigate its steady state creep behaviour at room temperature. Preliminary indentation tests on standard Al and Pb samples confirmed that the procedures employed gave results consistent with published data from conventional creep tests. Indentation tests on the binder gave a stress exponent consistent with a viscous flow mechanism for steady state creep. Comparisons are presented between the macroscopic creep rates exhibited by the composite over a range of temperature, with standard fixed load conditions, and predictions based on the binder indentation data, using a simple geometrical model for the constituent geometry in the composite. In general, good agreement is exhibited.

INTRODUCTION

Structure of Barrikade®

A novel fire-resistant composite material has recently been developed by TWI, Granta Park, Cambridge, UK, and has been given the trade name of Barrikade®. This composite is made of particles of a mineral, vermiculite, with, in the standard product, sodium silicate present as a binder. Vermiculite is a layered mineral, similar to mica, which has a marked capacity to undergo exfoliation. When rapidly heated, vaporisation of interlayer water causes pronounced expansion normal to the layers [1], resulting in highly anisotropic grains with a "concertina-like" structure. The vermiculite particles used in Barrikade® are typically 4-8 mm in diameter. These are mixed with the binder in aqueous solution before drying at 100°C. The final structure is a coarse agglomerate of particles, held together by relatively small bonds at contact points. This is illustrated in figure 1, which shows sections of polished specimens in which the pores have been infiltrated with resin. This material has a relatively low density (~300 kg m-3 [2]), and the limited area of bonding is expected to promote low strength and stiffness.

The binder may also be mixed with 3.1vol% of zinc oxide, in order to promote the water resistance of the final composite, by causing a chemical curing process in the binder.

The material is fire-resistant and produces no toxic emissions when subjected to heat or flame, but some properties remain unknown. Many applications, such as a fire door core material, will require use at elevated temperatures, and knowledge of the creep behaviour is of importance.

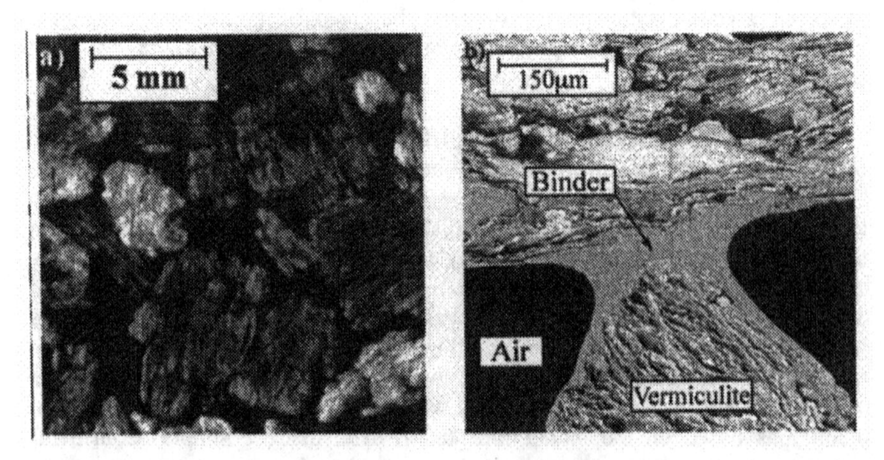

Figure 1. a) Low magnification optical micrograph of a mounted and polished cross section of Barrikade® and b) section through a typical bond between two vermiculite particles, imaged in the SEM using backscattered mode.

Nanoindentation Measurements of Creep

Conventional fixed load creep tests may be used to investigate the creep behaviour of the whole composite, but data for the constituents is required to deduce why the composite behaves in this way. Due to the dehydration step it is difficult to produce sodium silicate samples with a large cross section, meaning that thin film techniques must be used to obtain binder properties. A common method used to investigate the mechanical properties of thin films is nanoindentation [3]. In this technique creep is often a problem when collecting data [4], but recent investigations have looked at using this procedure to quantitatively examine creep.

The initial work on which these techniques are based came from the hot hardness tests of Tabor *et al* [5], results from which have been applied to nanoindentation. Several techniques have been developed, including indentation load relaxation (ILR) [6], constant rate of loading (CRL) [7] and the procedure that is most appropriate for use with our equipment, the constant load test devised by Mayo *et al* [8]. The method uses a dwell at constant load and measures how the indented depth, h, changes. In analysing the data the stress is approximated by load over contact area, and the strain rate by:

$$\dot{\varepsilon} = \frac{1}{h}\frac{dh}{dt} \tag{1}$$

EXPERIMENTAL

Creep Parameters

In this investigation the creep has been examined during the steady state creep region. The strain rate observed in this regime is commonly described by:

$$\dot{\varepsilon} = C\sigma^n \exp\left(-\frac{Q}{RT}\right) \tag{2}$$

Where C is a constant, σ is the applied stress (Pa), R is the gas constant and T is the absolute temperature (K). The steady state creep behaviour of the material is defined by the stress exponent, n, and the activation energy for creep, Q, which describe the effect on creep rate of stress and temperature respectively. For the temperature invariant case the equation simplifies to:

$$\dot{\varepsilon} = A\sigma^n \tag{3}$$

Where A is a constant and other terms are as defined previously. In this work creep data has been analysed with the above equations, to determine values for n and, where possible, Q.

Investigation of the Creep of Barrikade®

The creep behaviour of the composite as a whole and that of the individual constituents was examined. Samples of Barrikade® both with and without the standard addition of 3.1vol% ZnO to the binder were investigated, as the chemical curing effect of this addition was expected to be beneficial in resisting creep.

The creep behaviour of samples of Barrikade® (in the form of 30 mm side length cubes) was examined under a range of compressive stresses (70 kPa to 250 kPa) and temperatures (room temperature to 500 °C) in conventional static load tests. For comparison smaller samples (25mm×5mm×5mm) were tested in a push rod dilatometer. The alumina push rod, which in thermal expansivity measurements allows the change in linear dimension of a sample to be recorded, was used to apply a small load to the sample and also measure the strain rate. Stresses applied were from 2 to 7 kPa and temperatures from 100 to 500°C. In both tests measurements were carried out for between 6.5×10^4 and 13×10^4 seconds, until steady state was clearly seen.

As the tests are carried out for different stresses and at different temperatures, values for C, n and Q (from equation (2)) may be determined. Plotting $\log_{10}$ of the strain rate against $\log_{10}$ of the stress should, for steady state creep, produce a straight line with gradient equal to n. If the natural log of the strain rate is plotted against the reciprocal of the absolute temperature, a straight line should once again be produced, where the gradient is the negative of the activation energy over the ideal gas constant.

Investigation of the Creep of Vermiculite

The dilatometer was also used to examine the creep behaviour of individual particles of vermiculite, both in the through thickness and in plane directions. This material can only be obtained in small samples (<10mm diameter), and is very weak and compliant, needing very low loads to avoid crushing.

Investigation of the Creep of Sodium Silicate Binder

In order to examine the creep of the binder phase, nanoindentation was used. Prior to making measurements on binder samples, the creep properties of samples of commercially pure lead (Alfa Aesar) and aluminium (Goodfellow) were investigated. Polished samples were mounted in a nanoindenter (Micro Materials NanoTest), and indented with a range of maximum loads between 0.5 and 10 mN. The loading and unloading rate was chosen to give a total loading time of 20 seconds. Once the maximum load was reached the load was held constant, usually for a period of 1200 seconds, although 3000 seconds was required for some of the more creep prone samples. During this stage the depth of the indenter recorded with time. A 600 second dwell at partial unload was used to determine the thermal drift occurring during the experiment. The curves obtained were then analysed using the method of Mayo *et al* [8], to give the strain rates at different stresses. Samples of binder were made up, both with and without the standard addition of 3.1vol% zinc oxide. The samples were approximately 100-200μm thick, and were indented using the same program as the lead and aluminium samples. All measurements were performed at 25°C.

RESULTS AND DISCUSSION

The Creep of Barrikade®

Figure 2 shows examples of the strain recorded with time for both test methods. The graphs used to determine n and Q are shown in figure 3, and the values themselves are given in table I.

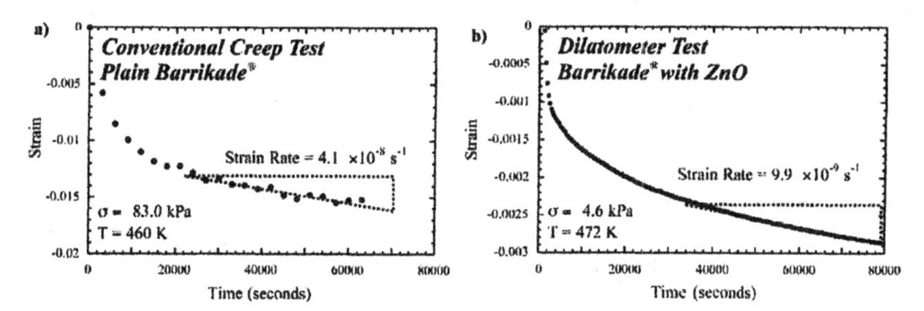

Figure 2. Experimental results for a) the conventional creep testing of plain Barrikade®, and b) dilatometer creep testing of Barrikade® with ZnO.

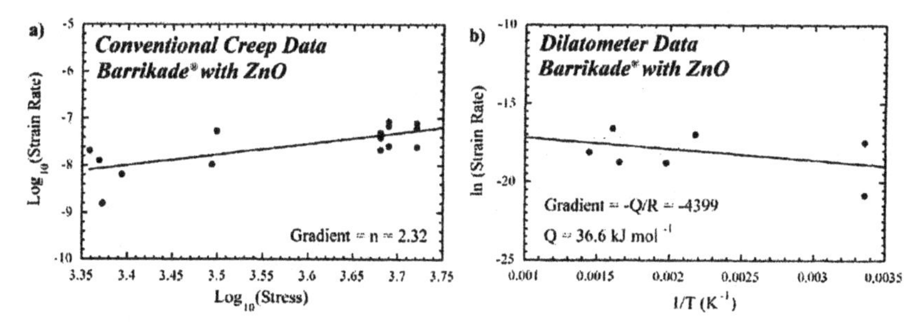

Figure 3. Examples of the analysis used to obtain values of a) the stress exponent, n, and b) the activation energy for creep, Q, both for Barrikade® with ZnO.

Table I - Values for creep parameters of Barrikade® determined in conventional creep testing and using the dilatometer.

Material	Test	C	n	Q (kJ mol^{-1})
Barrikade® with ZnO	Conventional	1.3×10^{-15}	2.0	36.6
	Dilatometer	1.0×10^{-15}	2.3	8.2
Plain Barrikade®	Conventional	2.2×10^{-11}	0.7	5.9
	Dilatometer	6.0×10^{-12}	1.2	5.2

The results show good agreement between the two techniques with most of the parameters giving similar values when measured in conventional tests and the dilatometer. The exception is in Q for Barrikade® with ZnO in the binder. As can be seen in figure 3 the graphs have a degree of scatter, and this could lead to inaccuracy in the results. The addition of ZnO to the binder can clearly be seen to have an effect on the behaviour of the composite, changing the values of C, n and Q. The lower value of C reflects the reduced tendency of the composite with ZnO to creep, probably because of the chemical curing effect. The higher value of n indicates that the creep rate of the material with zinc oxide is more stress sensitive than the plain material. The values of Q are generally low compared to activation energies for many processes, indicating that the creep behaviour of Barrikade® is very temperature sensitive.

The Creep of Sodium Silicate

Calibration using lead and aluminium

The results for the dwell period for samples of lead and aluminium are shown in figure 4, along with the lines used to calculate n.

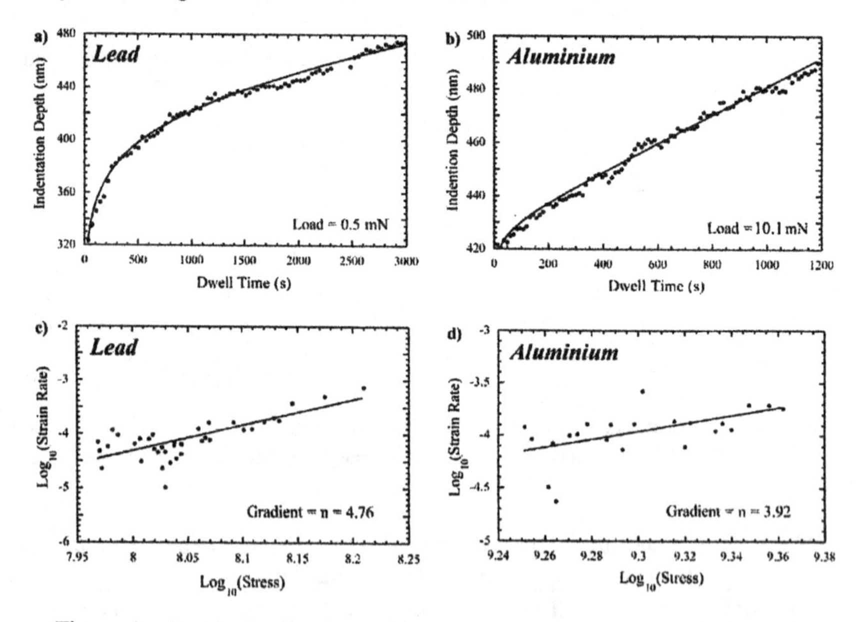

Figure 4. Load dwell data for nanoindentation creep testing of a) commercially pure lead, and b) commercially pure aluminium, and the analysis to obtain the value of the stress exponent, n, for c) lead and d) aluminium.

A relatively large degree of scatter is seen in the results, so linear regression lines of best fit have been used to determine the gradient. The scatter probably arises from the roughness on the experimentally obtained traces.

The values calculated for lead and aluminium may be compared to the literature values [9]. For lead we obtained n = 4.76, compared with a literature value of 5.0, and for aluminium we found n = 3.92, compared with 4.4. These results suggest that there is some parallel between the creep properties investigated by nanoindentation and conventional methods, and indicate that the values found in the tests on the binder may relate quite well to macroscopic properties. Similar parallels between nanoindentation values for n and those of macroscopic tests have been found previously [8], [10].

Measurements on the binder phase

As the nanoindentation tests were only carried out at room temperature, it was not possible to find a value for Q. Instead the data were analysed with equation (3). Examples of the results for the dwell period and the graphs used to calculate the values of n are shown in figure 5, and the values of n and A determined are given in table II.

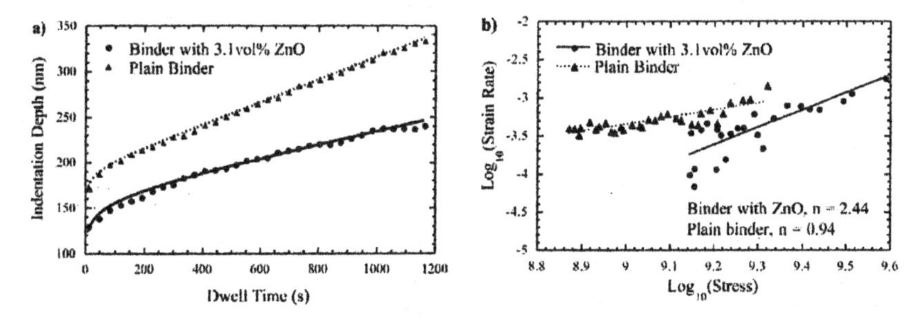

Figure 5. a) Load dwell data for nanoindentation creep testing of the binder phase with and without the addition of 3.1vol% ZnO, and b) the analysis to obtain the value of the stress exponent, n.

Table II - Average values for binder creep parameters determined in nanoindentation.

Material	A	n
Plain Binder	5.22×10^{-13}	1.09
Binder with ZnO	1.51×10^{-13}	2.44

The results indicate a substantial difference between binder with and without an addition of 3.1vol% ZnO. As predicted, the presence of ZnO reduces the creep rate. The value of n for plain binder close to 1 may be significant, as this is a characteristic of creep by viscous flow. This could reflect the binder not being fully dehydrated, and remaining as a viscous gel, rather than a glassy solid. When

ZnO is added the additional rigidity afforded by the chemical curing effect appears to change the mechanism.

The Creep of Vermiculite

The results from vermiculite in plane and through thickness from the conditions of highest stress and temperature are shown in figure 6 with Barrikade® from the same conditions for comparison. The changes in dimension are very small, close to the accuracy of the equipment, and therefore, when compared to the high creep rate of the binder, the contribution to creep by this constituent is negligible.

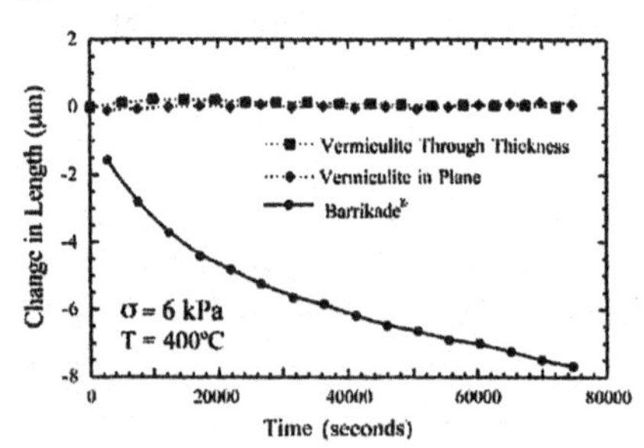

Figure 6. The results from testing vermiculite in the in plane and through thickness directions in the dilatometer. The results for Barrikade® for the same conditions are included for comparison.

Comparison of Composite Creep Rate with Predicted Values

As the creep behaviour of the constituents and the composite have been characterised, it should be possible to make a prediction of the composite creep rate and see if the values obtained are reasonable.

The situation is somewhat simplified by the assumption, justified by the results from figure 6, that the creep rate of the vermiculite in both the in plane and through thickness directions is negligible. Nevertheless, a full modelling treatment would have to consider effects such as rotation of the grains about the points of contact. For the current aim of determining the reliability of the creep data obtained from the nanoindenter, we have chosen the simpler route of using a slab model.

For the composite we can say the strain rate is given by:

$$\dot{\varepsilon}_c = \frac{dl_c}{dt} \cdot \frac{1}{L_c} \tag{3}$$

As vermiculite has a negligible creep rate:

$$dl_c = dl_b = \varepsilon_b L_b = \varepsilon_b L_c f_b \tag{4}$$

$$\dot{\varepsilon}_c = f_b \dot{\varepsilon}_b \tag{5}$$

Where ε is the strain, dl is the extension, L is the original length, f is the volume fraction, t is time and the subscripts c and b refer to the composite and binder respectively.

The volume fraction of the binder is calculated to be close to 10%, so this equation may be used to convert the binder data found in nanoindentation to creep rates for the composite under the same conditions. This is still not enough to allow comparison, as the nanoindentation data is collected at a much higher stress than the data for the composite. In order to make the comparison we need to use the experimentally determined value of n for the binder to predict how the slab model composite would behave at the stresses we have actually measured.

The comparison is shown in figure 7. A reasonable match is found between the measured and predicted room temperature creep rate for Barrikade® without ZnO. For material with the addition, the actual room temperature creep rate was below the sensitivity of the equipment used. This places an upper limit on the creep rate of 10^{-12} s^{-1}, which is consistent with the prediction from the nanoindentation results.

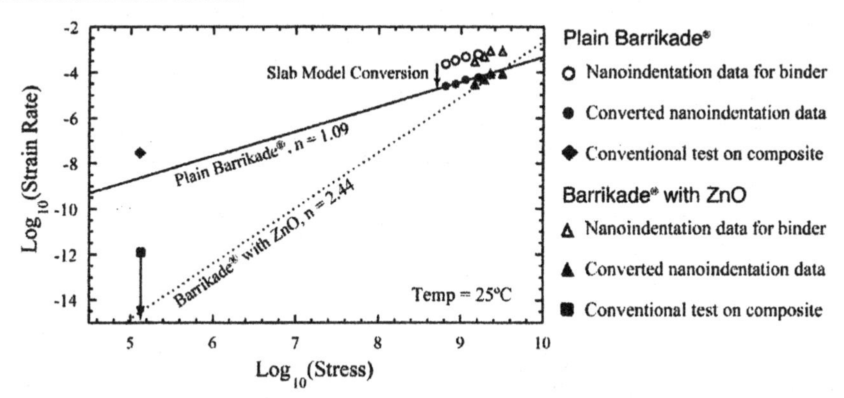

Figure 7. Comparison of nanoindentation results and conventional creep tests, using the slab model to convert the binder creep rate to the composite creep rate.

CONCLUSIONS

The tests on Barrikade® show that the material creeps at a relatively high rate, and is highly temperature sensitive. The addition of zinc oxide to the binder seems to reduce the creep rate, making this addition a favourable one.

It appears that the technique of nanoindentation creep measurements for determination of values of the stress exponent may be successfully applied to materials such as the sodium silicate binder, which are glassy solids or viscous gels. To the first approximation of the slab model the results agree quite well with the whole composite. These results give a good indication that if the binder is used without the addition of zinc oxide the creep process occurring is that of viscous flow.

FUTURE WORK

In order to complete the comparison of constituent and composite creep properties begun in this work it is necessary to obtain a measure of the activation energy for creep of the binder. This should be possible from nanoindentation by the use of a hot stage, to investigate the effect of temperature.

For the comparison to be improved, a more advanced model of the creep of the composite should be formulated, allowing for structural effects such as the rotation of the particles about their bonded points.

ACKNOWEDGEMENTS

Rgwould like to acknowledge the support of TWI and the EPSRC, through the PTP scheme.

REFERENCES

1. K.W. Cowling and J. White. "Vermiculite: Its Constitution, Properties and Uses" in *Int. Ceram. Congress*. Florence. (1954)

2. TWI, "Barrikade® Product Datasheet: Fire Resistant and Thermal Insulation Material". TWI: Cambridge.(2001)

3. W.C. Oliver and G.M. Pharr, "An Improved Technique for Determining Hardness and Elastic Modulus Using Load and Displacement Sensing Indentation Experiments." *J. Mater. Res.*, **7**(6) 1564-1583. (1992)

4. T. Chudoba and F. Richter, "Investigation of Creep Behaviour Under Load During Indentation Experiments and its Influence on Hardness and Modulus Results". *Surf. Coat. Techn.*, **148** 191-198. (2001)

5. A.G. Atkins, A. Silverio, and D. Tabor, "Indentation Hardness and the Creep of Solids". *J. Inst. Met*, **94** 369-378. (1966)

6. W.R. LaFontaine, B. Yost, R.D. Black, and C.Y. Li., "Indentation Load Relaxation Experiments with Indentation Depth in the Submicron Range". *J. Mater. Res.*, **5**(10) 2100-2106. (1990)

7. M.J. Mayo. and W.D. Nix, "A Micro-Indentation Study of Superplasticity in Pb, Sn, and Sn-38wt% Pb". *Acta Metall.*, **36**(8) 2183-2192. (1988)

8. M.J. Mayo, R.W. Siegel, and A. Narayansamy, "Mechanical Properties of Nanophase TiO_2 as Determined by Nanoindentation". *J. Mater. Res.*, **5**(5) 1073-1081. (1990)

9. H.J. Frost. and M.F. Ashby, "Deformation Mechanism Maps - The Plasticity and Creep of Metals and Ceramics". Oxford, Pergamon. (1982)

10. B.N. Lucas and W.C. Oliver, "Indentation Power-Law Creep of High-Purity Indium". *Metall. Mater. Trans.*, **30A** 601-610 (1999).

AN EXPERIMENTAL ASSESSMENT OF USING CRACK-OPENING DISPLACEMENTS TO DETERMINE INDENTATION TOUGHNESS FROM VICKERS INDENTS

J. J. Kruzic and R. O. Ritchie[*]
Materials Sciences Division
Lawrence Berkeley National Laboratory, and
Department of Materials Science and Engineering
University of California, Berkeley, CA 94720

ABSTRACT

Recently, a method for determining indentation fracture toughness from Vickers hardness indentations has been proposed which is based on a comparison of measured crack-opening displacements to computed values. To provide a first test of this method, experiments were conducted on a commercial silicon carbide ceramic, Hexaloy SA. Using the method, a toughness value within 10% of that measured using conventional fracture toughness testing was obtained; however, there was poor agreement between the measured and computed crack-opening profiles. Such discrepancies raise concerns about the suitability of this new method for determining the toughness of ceramics. Possible causes for these discrepancies are discussed, including subsurface lateral cracking and crack formation during loading.

INTRODUCTION

Indentation toughness testing is an attractive alternative to more costly fracture mechanics experiments for determining the fracture properties of brittle materials. The typical method used involves measuring the radial cracks emanating from Vickers diamond microhardness indents and applying the semi-empirical relationship:[1]

$$K_c = \chi\sqrt{\frac{E}{H}}\frac{P}{a^{\frac{3}{2}}} , \tag{1}$$

[*] Corresponding author. Tel: +1-510-486-5798; fax: +1-510-486-4881. *E-mail address:* roritchie@lbl.gov (R. O. Ritchie)

where P is the indentation load, E is Young's modulus, H is the Vickers hardness, a is the radial crack length measured from the center of the indent, and χ is an empirically determined "calibration" constant taken to be 0.016 ± 0.004.[1] One drawback of this method is that there is a large uncertainty (± 25%) in χ; consequently, the ability to obtain a precise value for the toughness is limited. Also in ceramics the value of H is not always invariant with respect to changing loads owing to indentation size effects, which gives further uncertainty to the K_c values obtained by this method (Eq. 1).[2] Finally, for materials that exhibit rising R-curve behavior due to crack bridging or transformation toughening, additional uncertainties arise due to the fact that indentation toughness essentially gives a random point on the R-curve.

A new, non-empirical, method utilizing newly available solutions for the crack-opening displacements of Vickers indentation cracks has recently been proposed by Fett for determining the indentation toughness.[3] With this method, the crack-opening profile, $u(r)$, for a linear-elastic Vickers indent crack is expressed in terms of the near-tip stress intensity, K_{tip}, such that:[3]

$$u(r) = \frac{4K_{tip}\sqrt{b}}{\pi E'}\left(A\sqrt{1-\frac{r}{a}} + B\left(1-\frac{r}{a}\right)^{\frac{3}{2}} + C\left(1-\frac{r}{a}\right)^{\frac{5}{2}}\right), \quad (2)$$

where E' is the plane strain modulus (i.e., $E' = E/1-\nu^2$, where ν is Poisson's ratio) while r, b, and a are the radial position, contact-zone radius, and the crack length, respectively, as measured from the center of the indent. The coefficients A, B, and C are given by:[3]

$$A = \sqrt{\frac{\pi a}{2b}}\ , \quad (3)$$

$$B \cong 0.011 + 1.8197\ln\left(\frac{a}{b}\right)\ , \quad (4)$$

$$C \cong -0.6513 + 2.121\ln\left(\frac{a}{b}\right)\ . \quad (5)$$

Simplifying the first term in Eq. 2 gives the Irwin near-tip crack opening solution:

$$u(r) = \frac{K_{tip}}{E'}\sqrt{\frac{8(a-r)}{\pi}}\ . \quad (6)$$

For the simple case of an ideally brittle, non-toughened ceramic, Eqs. 2-5 provide a means for deducing the intrinsic (crack initiation) toughness, K_o, from the measured crack-opening profile. One necessary assumption, as in other indentation toughness methods, is that cracking occurs due to the residual tensile stresses during unloading and accordingly crack arrest occurs when the near-tip stress intensity $K_{tip} = K_o$.

This method also has potential for use with bridging ceramics, such as grain-elongated Si_3N_4 and ABC-SiC.[4-6] By separating out the bridging displacements, which distort the crack-opening profile, from that expected in the traction-free case (i.e., that given by Eq. 2-5)[3], the intrinsic toughness, or crack-initiation point on the R-curve, may be determined, as well as values for the bridging contribution to the toughness.

With this goal in mind, the present objective is to asses the suitability of Fett's method for assessing fracture toughness.[3] A commercial silicon carbide, Hexoloy SA, was chosen for this study for several reasons. Firstly, Hexoloy SA fractures transgranularly without crack bridging. Consequently, there is a single-valued fracture toughness for this material, which is independent of crack length. i.e., no R-curve behavior; moreover, the toughness of this material has been well characterized.[4,6] Additionally, a well defined radial crack system, necessary for this study, can be readily produced by indenting silicon carbide ceramics.[1,7] Finally, as silicon carbides can be readily produced to develop grain bridging,[4,6] the study can be extended to ceramics with R-curve behavior using the same nominal material. In order to test the validity of this new method, results are compared to those obtained by both conventional fracture mechanics testing, using precracked specimens, and conventional indentation toughness methods, using Eq. 1.

EXPERIMENTAL PROCEDURES

Disks of a pressureless sintered silicon carbide, Hexoloy SA, were ground flat and lapped to a 1 μm finish using diamond compounds. A 4 kg load was used to produce Vickers indentations. While using a range of loads would be ideal for this assessment, lower loads produced too short of indent cracks while larger loads caused surface chipping. Accordingly, a 4 kg indentation load allowed the maximum radial crack lengths without chipping the sample surface. Three methods for determining indentation toughness were compared:

1. *Crack-opening profile (COP) method:* Using the proposed method of Fett,[3] denoted here as the COP method, the intrinsic toughness, K_0, was determined by measuring crack-opening profiles, $u(r)$, and computing the value of K_0 using Eqs. 2-5. Crack openings were measured in a field-emission scanning electron microscope (FESEM), with a maximum obtained resolution of 5 nm for the full crack opening, $2u$. The optimal value of K_0 was determined to be that which produced the best least-squares fit of Eqs. 2-5 to the experimentally measured crack-opening profile.
2. *Near-tip (NT) method:* To assess whether the full crack-opening profile was needed for this method, a identical technique to the COP method was utilized *except* that only the near-tip crack-opening data were used. Accordingly, for this near-tip (NT) method, the Irwin solution (Eq. 6) was used instead of Eqs. 2-5; this method was carried out using 5, 10, 15, and 20 μm of crack-opening data.
3. *Traditional indentation toughness (TIT) method:* Additionally, the above results were compared to toughness value obtained using the standard indentation toughness procedures.[1] Specifically, the indent size and crack length were measured and the toughness computed using Eq. 1.

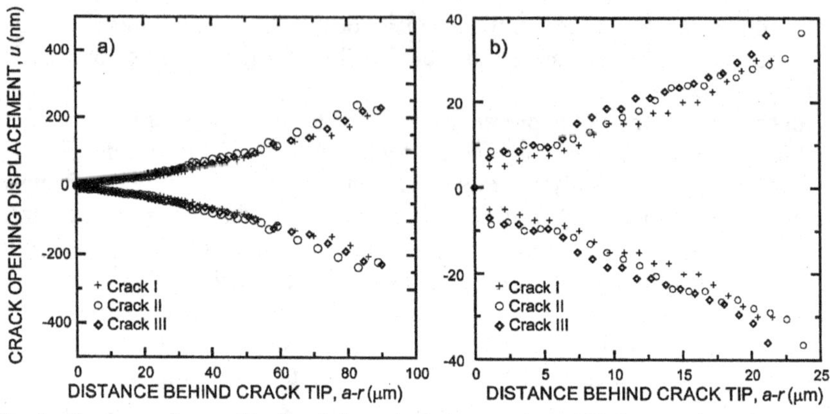

Fig. 1: Crack-opening profiles for a) the entire indent cracks and b) the near-tip regions.

Finally, all indentation toughness results were compared to values measured using *conventional fracture mechanics methods*, based on studies which used precracked disk-shaped compact-tension specimens.[5,7]

RESULTS

Fig. 1a shows the full crack-opening profiles measured for three Vickers indent cracks, while Fig. 1b shows a close up of the near-tip region for each crack. The three cracks are denoted I, II, and III. Fig. 2 shows micrographs of cracks I and II along with a cross section that reveals the nature of the subsurface cracking which occurs in this material. Estimates of the intrinsic toughness for Hexoloy SA, K_o, were made using the results from Fig. 1 along with Eqs. 2-5, i.e., using the COP method. The deduced K_o values were 2.0 MPa$\sqrt{m}$ for crack I and 2.3

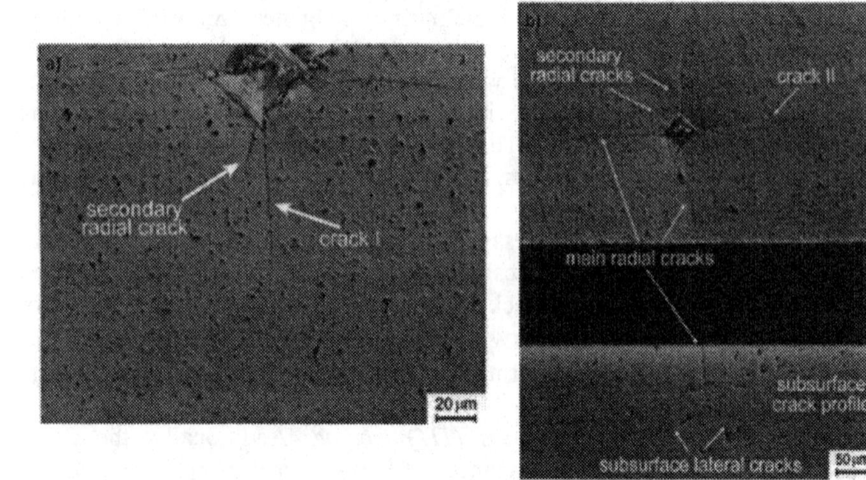

Fig. 2: Micrograph showing a) crack I and b) crack II along with a subsurface crack profile

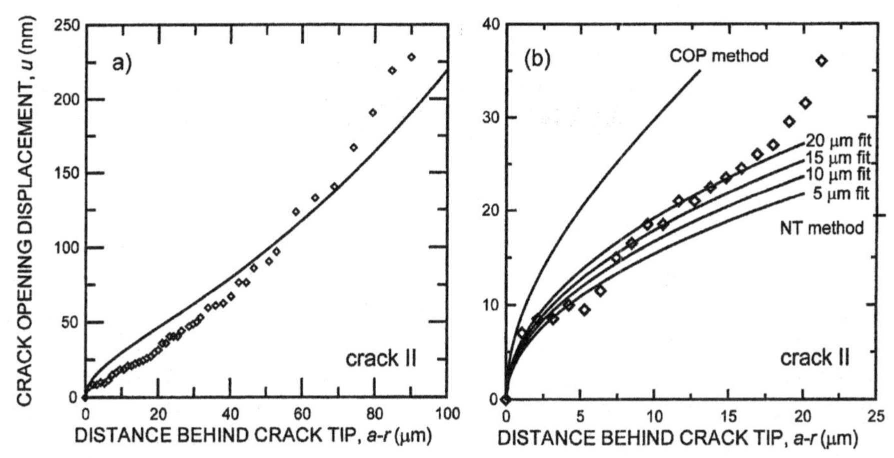

Fig. 3: Part a) shows the computed crack opening for crack II based on the COP method along with the full measured crack-opening profile while b) shows the near-tip region along with the computed openings based on the NT and COP methods.

MPa√m for cracks II and III. Using these deduced toughness values, the crack-opening profiles were computed for each crack and compared to the measured crack openings. The results looked similar for each case, with Fig. 3a showing the predicted crack opening for crack II using K_0 = 2.3 MPa√m. The above results obtained using the COP method are tabulated in Table I along with the results from the NT method. The later results were obtained using only the near-tip data (Fig. 1b) with the Irwin crack-opening solution (Eq. 6), and were found to be dependent on the amount of data used for the fit (Table I). As for the COP method, the best fit near-tip crack profiles were computed for each case. Again these results were similar for each crack; one example set of results, for crack II, are shown in Fig. 3b.

Table I: Fracture toughness results using crack-opening profile measurements (COP and NT methods)

	Crack I	Crack II	Crack III
COP method (Eqs. 2-5)	2.0 MPa√m	2.3 MPa√m	2.3 MPa√m
NT method (Eq. 6) 5 μm fit	0.93 MPa√m	1.3 MPa√m	1.4 MPa√m
NT method (Eq. 6) 10 μm fit	1.0 MPa√m	1.4 MPa√m	1.2 MPa√m
NT method (Eq. 6) 15 μm fit	1.2 MPa√m	1.5 MPa√m	1.4 MPa√m
NT method (Eq. 6) 20 μm fit	1.3 MPa√m	1.6 MPa√m	1.5 MPa√m

Finally, to compare the indentation toughness results obtained from the crack-opening profile measurements to those obtained by the tradition indentation toughness method,[1] measurements were made for cracks at four indents, giving an average indentation toughness (TIT method) of 2.1 ± 0.7 MPa√m for Hexoloy SA.

DISCUSSION

The results obtained in the present study using the COP method compare favorably with the typical reported values of roughly 2.5 MPa√m obtained using conventional fracture mechanics testing with disk-shaped compact-tension specimens.[4,6] Specifically, for cracks II and III, the deduced value of 2.3 MPa√m falls within 10% of typical reported values. For crack I, a slightly lower value (2.0 MPa√m) was obtained; however, Fig. 2a reveals that there was a significant secondary radial crack near the main crack I. Such secondary cracking may have relieved some of the residual stress which holds the crack open, thus affecting the crack-opening profile and the deduced toughness values for crack I. Cracks II and III were specifically chosen such that no (crack II) or minimal (crack III) secondary cracking could be observed near the main crack. Thus, considering only the results for cracks II and III, it may be concluded that reasonable estimates of the intrinsic toughness of Hexoloy SA may be obtained using the COP method if the cracks analyzed are specially chosen such that secondary radial cracking is at a minimum.

One major disadvantage of the semi-empirical traditional indentation toughness (TIT) method is the large scatter in the results; indeed, in the present study a wide range of values from 1.4 to 2.8 MPa√m was obtained for Hexoloy SA. Although this range overlaps with the typical reported values of ~2.5 MPa√m obtained from conventional fracture mechanics methods,[4,6] the lowest values of the range differ by some 45%. Conversely, the COP method produced values within 10% of the expected value. Furthermore, the deduced toughness value, K_o, was still within 20% for the case of crack I, where secondary radial cracking is believed to have affected the results. Thus, these first results indicate there is definite promise of using crack-opening displacement measurements to deduce toughness values for ceramics.

There are serious concerns about using this method, however, based on the large discrepancies between the computed and measured crack-opening profiles seen in Fig. 3. The results in Fig. 3 were typical for all three cracks in that the crack-opening shape did not agree with that predicted by Eqs. 2-5, particularly in the near-tip region. Toughness results obtained using the NT method further illustrate this point. Fitting the Irwin solution (Eq. 6) to the near-tip crack-opening data yielded toughness values up to 50% lower than that obtained using the COP method. Although the COP method takes the entire crack opening into account, and accordingly should give more accurate results, such large discrepancies were not expected. These results indicate that the actual near-tip crack openings are much smaller than would be expected for a crack held open just below the condition for critical fracture, i.e., $K_{tip} = K_o$. It is important to note that Hexoloy SA fractures transgranularly, so the smaller than expected crack openings may not be attributed to crack-face interactions such as bridging.

The observed discrepancies may be attributed to several possible factors. The first of these is the presence of subsurface lateral cracking, as shown in Fig. 2b. While subsurface lateral cracks are not uncommon during Vickers indentation,[8-10] the effects of such cracking have largely been ignored in the analysis methods for indentation toughness. For the traditional indentation toughness (TIT) method, i.e., Eq. 1, lateral cracking is not considered;[1] however, it is likely to account for some of the uncertainty in the empirical calibration factor. One effect of lateral cracking is to alter the residual stress field, which in turn affects the crack-opening profiles and hence the toughness values deduced by the COP and NT methods. Specifically, relaxation of residual stresses would permit the cracks to close more than expected or have a different opening shape than expected; indeed, both of these scenarios are observed in the results of Fig. 3. Accordingly, it is expected that future developments in this method will need to account for such lateral cracking when predicting crack-opening shapes.

Finally, the basic assumption of all current indentation toughness methods, that cracking occurs during unloading as a result of the residual tensile stresses, has been brought into question by the work of Cook and Pharr.[9] Specifically, their *in situ* optical microscopy studies of transparent materials showed that while some glasses exhibit the expected behavior of cracking during unloading, several transparent ceramic materials in fact exhibited cracking immediately upon loading the sample. Such results raise questions about the general applicability of all indentation toughness methods, including the TIT method which has now been used for more than twenty years. It is currently unknown if cracking occurs upon loading or unloading in SiC; however, the possibility of cracking during the loading portion of indentation may contribute to the observed discrepancies between predicted and measured crack openings seen in Fig. 3.

In summary, it is clear that there are significant uncertainties underlying a general indentation toughness test method for ceramics based on the measurement of crack-opening displacements. While the method as applied here may work well for some ceramic materials, further experiments for a range of brittle materials are needed. Furthermore, for silicon carbide, it is apparent that a more through understanding of the mechanisms and interactions of Vickers indent cracking is needed before such a method may be suitably applied for reliable toughness determination. Thus, while the crack-opening profile (COP) indentation toughness method proposed by Fett[3] indeed holds promise, there is a need for further investigation in this area.

CONCLUSIONS

Based on an experimental study to assess the use of crack-opening displacements to determine the fracture toughness of a commercial silicon carbide ceramic, Hexoloy SA, from Vickers indentation cracks, the following conclusions are made:

1. An intrinsic toughness value of $K_o = 2.3$ MPa$\sqrt{m}$, which was within 10% of values reported using standard fracture mechanics specimens, was obtained by fitting the expected full crack-opening profile to the measured openings. Such close correspondence illustrates the potential of such methods.
2. The deduced toughness was lower (2.0 MPa$\sqrt{m}$) for an indentation crack where significant secondary radial cracking was evident. In this case, secondary radial cracking was believed to affect the results by reliving some of the residual stresses, resulting in smaller crack openings and a smaller deduced toughness value.
3. In all cases, there were significant discrepancies between the measured and computed crack openings, even in the absence of secondary radial cracking. Possible explanations for these discrepancies include subsurface lateral cracking, which was directly observed and can affect the residual stress field, and the possibility of cracking during the loading portion of the indentation process.
4. Although it appears that methods using the crack-opening displacements to determine the fracture toughness of ceramics from Vickers indents hold promise, a more thorough understanding of the complexities of indent cracking is needed for this to become a reliable and accurate test method.

ACKNOWLEDGEMENTS

Work supported by the Director, Office of Science, Office of Basic Energy Sciences, Division of Materials Sciences and Engineering of the U.S. Department of Energy under Contract No. DE-AC03-76SF00098.

REFERENCES

[1]G. R. Anstis, P. Chantikul, B. R. Lawn and D. B. Marshall, "A critical evaluation of indentation techniques for measuring fracture toughness. I. Direct crack measurements," *J. Am. Ceram. Soc.,* **64** [9] 533-8 (1981).

[2]J. Gong, J. Wang and Z. Guan, "Indentation toughness of ceramics: A modified approach," *J. Mater. Sci.,* **37** [4] 865-9 (2002).

[3]T. Fett, "Computation of the crack opening displacements for Vickers indentation cracks," Report FZKA 6757, Forschungszentrum Karlsruhe, Karlsruhe, Germany, 2002.

[4]C. J. Gilbert, J. J. Cao, W. J. MoberlyChan, L. C. De Jonghe and R. O. Ritchie, "Cyclic fatigue and resistance-curve behavior of an *in situ* toughened silicon carbide with Al-B-C additions," *Acta Metall. Mater.,* **44** [8] 3199-214 (1996).

[5]C.-W. Li, D.-J. Lee and S.-C. Lui, "R-curve behavior and strength for in-situ reinforced silicon nitrides with different microstructures," *J. Am. Ceram. Soc.,* **75** [7] 1777-85 (1992).

[6]C. J. Gilbert, J. J. Cao, L. C. De Jonghe and R. O. Ritchie, "Crack-growth resistance-curve behavior in silicon carbide: small versus long cracks," *J. Am. Ceram. Soc.,* **80** [9] 2253-61 (1997).
[7]J. J. Petrovic and L. A. Jacobson, "Controlled surface flaws in hot-pressed SiC," *J. Am. Ceram. Soc.,* **59** [1-2] 34-37 (1976).
[8]B. R. Lawn and M. V. Swain, "Microfracture beneath point indentations in brittle solids," *J. Mater. Sci.,* **10** [1] 113-22 (1975).
[9]R. F. Cook and G. M. Pharr, "Direct observation and analysis of indentation cracking in glasses and ceramics," *J. Am. Ceram. Soc.,* **73** [4] 787-817 (1990).
[10]D. B. Marshall, B. R. Lawn and A. G. Evans, "Elastic/plastic indentation damage in ceramics: the lateral crack system," *J. Am. Ceram. Soc.,* **65** [11] 561-6 (1982).

COMBINED NANOINDENTATION AND ACOUSTIC DETERMINATION OF THE ELASTIC PROPERTIES OF FLOAT GLASS SURFACE

Oriel Goodman and Brian Derby
The University of Manchester and UMIST
Manchester Materials Science Centre
Grosvenor Street
Manchester
M1 7HS
UK

ABSTRACT

Nanoindentation and scanning acoustic microscopy (SAM) have been used to characterise the mechanical properties of the two surfaces (air and tin-facing) of 4mm thick float glass. Both techniques can be used together to determine both the Young's modulus and Poisson ratio of the material.

It was found that the values for Young's modulus for the tin surface were 5% lower than those for the air surface. This discrepancy was believed to be due to variations in near-surface composition and surface damage between the two sides. It was concluded that differences found between the nanoindentation and SAM determined values of mean Young's modulus, for the surfaces of the glass, were caused by employing an assumed Poisson's ratio.

INTRODUCTION

The float glass process has been widely used since the 1960s, replacing the production of sheet and plate glass, and has been described in detail in various publications [1,2]. The advantage of this method is that the glass can be manufactured in a single operation and requires no further processing for a flat optically transparent finish. The two glass surfaces undergo different conditions during the process. One surface (air surface) is in contact with air throughout. The second (tin surface) is in contact with the tin bath and is then supported on a series of rollers in the annealing lehr. These different environments during processing may lead to differences in mechanical properties of the two surfaces. Here we use nanoindentation and acoustic microscopy to explore these differences.

NANOINDENTATION

Nanoindentation applies very small loads (typically <100mN) to a sharp diamond indenter of known geometry in order to make a permanent indentation in the surface of the test specimen. The three-sided Berkovich pyramid indenter is typically used for nanoindentation, as it is easier to produce with a sharp tip than the four-sided Vickers indenter. In addition, a Berkovich tip is straightforward to calibrate with no major flaws and is easy to calibrate [3]. As the indenter is pressed deeper into the surface of the sample measurement of this change in displacement is recorded as a function of the load. Assuming a perfectly manufactured Berkovich tip with a face angle 65.3° the area depth relation for the indenter is

$$A = 24.56h^2 \tag{1}$$

Where A is the indenter contact area and h is the calculated plastic depth.

The area can be determined from stiffness data using the relationship for the contact of isotropic elastic half-space by rigid indenters of various geometries as developed by Love and Sneddon [4-7].

$$S = \beta \frac{2}{\sqrt{\pi}} \sqrt{A} E_r \tag{2}$$

E_r is the reduced modulus that has the following relationship to the Young's modulus, E, and Poisson's ratio, ν, of the material.

$$\frac{1}{E_r} = \frac{(1-\nu^2)}{E} + \frac{(1-\nu_i^2)}{E_i} \tag{3}$$

Where the subscript i, refers to the properties of the indenter. The materials' Young's modulus can then be calculated providing the indenter contact area A, the indenter shape constant β (1.034 for a Berkovich tip) and a value for the stiffness of the contact S, (which can be calculated from the gradient of the unloading curve) are known.

In practice this model needs modification in order to allow for the effects of non-rigid indenters and elastic-plastic specimens as well as any compliance of the load frame and the indenter, which may take place during testing, as observed by Oliver and Pharr [8].

$$C = C_f + \frac{\sqrt{\pi}}{2E_r}\frac{1}{\sqrt{A}} \tag{4}$$

Where the overall compliance of the system C, is the reciprocal of the stiffness (i.e. the reciprocal of Eq. (2)), C_f is the frame compliance, which in addition to the specimen compliance, C_s gives the overall value C. However, these equations rely on the indenter having a perfect geometry and do not take into account of any imperfections that may occur due to the mechanical polishing of the indenter or any subsequent blunting during experimental work. The indenter therefore needs to be calibrated as described by Oliver and Pharr [8].

A further consideration has been proposed by Hay *et al.* [9], who discovered that only incompressible materials (i.e. those with a Poisson's ratio of 0.5) deformed to the shape of the indenter during indentation. It was consequently discovered that this difference in contact area for compressible materials could be directly related to the Poisson's ratio of the sample and the geometry of the indenter [9].

This correction factor γ, can then be included in Eq. (2) to allow for this discrepancy.

$$S = \gamma \frac{2}{\sqrt{\pi}}\sqrt{A}E_r \tag{5}$$

It is essential that the tip geometry be carefully calibrated before testing begins using a specimen of known Young's modulus and Poisson's ratio, in order to achieve more accurate and repeatable results.

ACOUSTIC MICROSCOPY

Acoustic methods are now widely used as a non-destructive method of characterising materials; in particular the scanning acoustic microscope (SAM) has become a particularly interesting analysis and micro characterisation tool. SAM allows structures within the material to be observed based on the knowledge that most solids have a very much lower level of acoustic absorption compared to optical absorption therefore, the material can be regarded as transparent to ultrasonic waves. SAM utilises very high frequency ultrasonic waves in a convergent beam to obtain a resolution of the order of a micrometre for depths of about a millimetre[10].

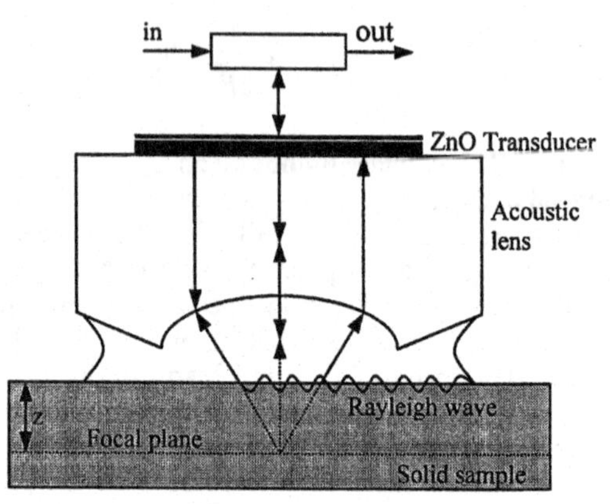

Fig. 1. Experimental configuration of the line focus acoustic microscope

The lens of the acoustic microscope consists of a sapphire disc, which has one flat polished face, and one face that has had a concave spherical surface ground into it. From the diagram of the acoustic transducer and lens (Fig. 1) it is apparent that there is a layer of water between the lens and the reflecting object or specimen. This acts as a coupling fluid between the two as at high frequencies the acoustic impedance mismatch between sapphire and air will not allow significant transmission of acoustic energy. The coupling fluid is needed to transmit the acoustical signal from the lens to the sample [11]. The attenuation of the coupling fluid is of vital importance as it determines the highest frequency and the shortest wavelength that can be used and consequently, the optimum resolution that can be achieved. Water is most commonly used as the coupling fluid [12] producing bulk shear and longitudinal waves as well as surface waves. The ZnO transducer in the SAM is used to transform a high-frequency radio frequency signal into an ultrasonic wave without changing the frequency. The sapphire lens then acts as an "acoustic delay line" [11] for the propagated wave.

The returned acoustic waves are received by a low-noise high-gain receiving amplifier [12], the signal is then displayed on an oscilloscope. The height of a chosen pulse, and hence the intensity of the echo from the specimen, is measured by a fast circuit, which retains this information as a dc signal until the next pulse is transmitted. This signal is stored as the intensity value at that particular position of the lens relative to the specimen. A complete image is then compiled by scanning the lens in a raster pattern, collecting the intensity of the echo at each position.

The strongest signal occurs when the specimen's surface is in focus. When the distance between the lens and the specimen is varied it is found that the contrast in an image changes even causing a reversal of the relative contrast. The response is an oscillating signal, which can be thought of as the material's acoustic signature, a typical response is shown in Fig. 2 as a V(z) curve. This curve represents the change in voltage V, which occurs when z, the distance between the specimen and the acoustic lens, is altered and the lens is defocused. This change in signal voltage results from interference between the reflected wave and specular waves generated in the specimen due to excitation by the pulsed wave [13].

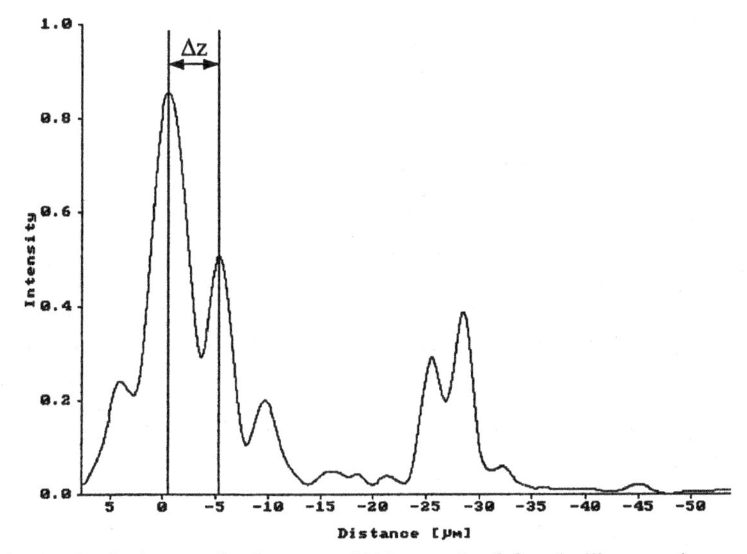

Fig. 2. Typical acoustic signature (V(z) curve) of fused silica specimen taken at 1GHz.

In Z-contrast, or Rayleigh wave, mode the sample is translated through focus in the Z direction, this process is often referred to as defocusing. Fig. 3 is a schematic illustration of the waves emitted and received to an acoustic lens. The majority of rays behave like ray aa' and are reflected specularly from the sample, they don't contribute significantly to the excitation of the transducer due to the angle at which they enter the lens. There are two waves of interest, the first travels straight to the specimen and is reflected back along the axis of the lens (ray bb'), the second wave (ray cc') is refracted at the Rayleigh angle θ_R, and excites a leaky Rayleigh wave in the surface of the specimen, which then radiates a wave back to the lens. The net signal received at the transducer is the phase-sensitive sum of these two waves [14].

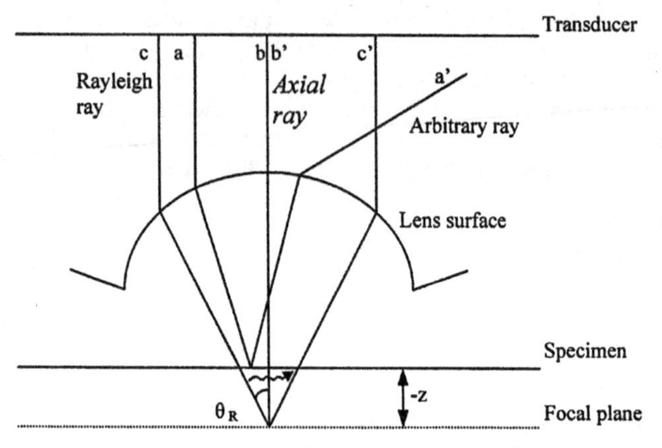

Fig. 3. Ray model of an acoustic lens with negative defocus: aa' is an arbitrary ray, which is reflected at such an angle that it misses the transducer (or else hits the transducer obliquely and therefore contributes little to the signal because of phase cancellation across the wave front); bb' is the axial ray, which goes straight down and returns along the same path; cc' is the symmetrical Rayleigh propagated wave, which returns to the transducer normally and so also contributes to the signal. The wavy arrow indicates the Rayleigh wave.

The period of the oscillations displayed in the acoustic signature is related to the Rayleigh wave velocity (v_R), on the surface of the material [12]. This velocity can then be used to characterise the surface elastic properties of the specimen as it is altered by the presence of surface roughness, surface breaking cracks and the presence of residual stress. The waves can also detect defects in the substructure of the sample for example subsurface cracks will result in a change in the ultrasonic signal. The Rayleigh wave velocity is related to the transverse (v_t) and longitudinal (v_l) acoustic wave speeds of the solid [15].

$$\left(\frac{v_R}{v_t}\right)^6 - 8\left(\frac{v_R}{v_t}\right)^4 + 8\left\{3 - 2\left(\frac{v_t}{v_l}\right)^2\right\}\left(\frac{v_R}{v_t}\right)^2 - 16\left\{1 - \left(\frac{v_t}{v_l}\right)^2\right\} = 0 \quad (6)$$

Kushibiki and Chubachi [16], described how the Rayleigh wave velocity can be determined from the distance between the interference maxima Δz, of the specimen's V(z) curve when applied to Eq. (7). Where v_o is the velocity of sound in the coupling fluid, θ_R is the critical angle for surface wave generation, f is the ultrasonic frequency and v_R is the Rayleigh wave velocity.

$$\Delta z = \frac{v_0}{2f(1-\cos\theta_R)}, \quad \sin\theta_R = \frac{v_0}{v_R} \tag{7}$$

Using these results, if the density (ρ) of the material is known, the surface Young's modulus (E_s) and the surface Poisson's ratio (ν_s) can be calculated using the following equations.

$$E_s = \frac{\rho v_t^2 (3v_l^2 - 4v_t^2)}{(v_l^2 - v_t^2)} \tag{8}$$

$$\frac{v_t}{v_l} = \frac{(1-2\nu_s)}{(2-2\nu_s)} \tag{9}$$

In order to detect the presence of these cracks, it will be necessary to obtain a Rayleigh wave velocity for a near perfect glass surface which should have a minimum number of surface breaking cracks, for example the upper surface of a sheet of float glass, to act as a reference point. This technique has been used previously to measure surface damage of various brittle materials [14,17-23].

The effect of surface-breaking cracks on the propagation of Rayleigh waves has been examined in a number of papers by Pecorari [20-23]. The variation of Rayleigh wave velocity was measured as the parameters controlling the cracks were altered, these included: the crack depth, the crack density and the crack orientation and it was concluded that the variation in SAW velocity caused by crack distributions was greater than caused by other forms of surface damage such as residual stress and roughness [21]. These observations were later incorporated into a theoretical model describing changes in Rayleigh wave velocity caused by the distribution of surface cracks[23]. However, in a later paper [22] Pecorari concedes that as residual stresses can result in the closure of some surface breaking cracks, this might lead to an underestimate of the surface damage as the Rayleigh waves will not be as dispersed. An extensive review and comparison of the numerous theoretical methods used to assess the effects of distributions of surface breaking cracks on the velocity of Rayleigh waves with experimentally determined results, is presented by Pecorari *et al.* [20] . They proposed that due to the difficultly in providing a clear quantitative description not reliant on statistical results, of surface damage in brittle materials produced from experimental data, there was a need to use "complementary experimental techniques".

EXPERIMENTAL

Samples of 4mm thick clear float glass were supplied by Pilkington Technology Centre for all of the experimental work carried out during this investigation. Nanoindentation was undertaken using a MTS XP Nano Indenter (MTS Systems Corporation, Minneapolis, USA) with a diamond indenter of Berkovich pyramid geometry. The load resolution of the system is 50nN and the displacement resolution is less than 0.02nm. The nanoindenter was calibrated using a sample of fused silica that had known values of Young's modulus and Poisson's ratio, which was indented to a maximum depth of 2μm. This revealed that a maximum depth of 2μm was sufficient to profile the surface and near-surface properties of glass.

In order to obtain continuous contact stiffness measurements as the indenter was driven into the sample, the continuous stiffness measurement (CSM) option was used. This technique measures the contact stiffness at many points along the loading curve, differing from the conventional load-displacement-time method where only one contact stiffness measurement is made from the unloading portion of the experiment at P_{max}. Hence, one indentation experiment can be used to provide all of the information that would usually take several load-displacement-time experiments with various peak loads [24] The loading cycle began with an initial indentation to measure the thermal stability of the specimen. Once this was below $0.05 nms^{-1}$ the indenter pulsed downwards into the specimen to a maximum depth of 2μm at thirty manually selected locations. For every indent, the indenter gave fifty results at intervals between the surface and the maximum depth. A constant strain rate of $0.05s^{-1}$ was used during loading.

The acoustic measurements were taken with a KSI 2000 Scanning Acoustic Microscope (Kramer Scientific Instruments GmbH, Herborn, Germany) with a 1GHz spherical lens and a ZnO transducer. Deionised water, which was maintained at room temperature throughout the testing, was used as the coupling fluid between the lens and the specimen. The SAM was calibrated by taking several V(z) curves of a sample of fused silica with known properties, from which the average distance Δz was used in Eq. (7). V(z) curves were also taken for the air and tin surfaces of specimens of 4mm thick float glass, and the Δz results were used to discover the v_R results for the surfaces. Two techniques for measuring V(z) were employed, the first required all of the measurements to be taken from the same spot on the sample with the lens being brought completely off the surface after each reading, in the second method the V(z) measurements were made at an array of locations.

RESULTS AND DISCUSSION

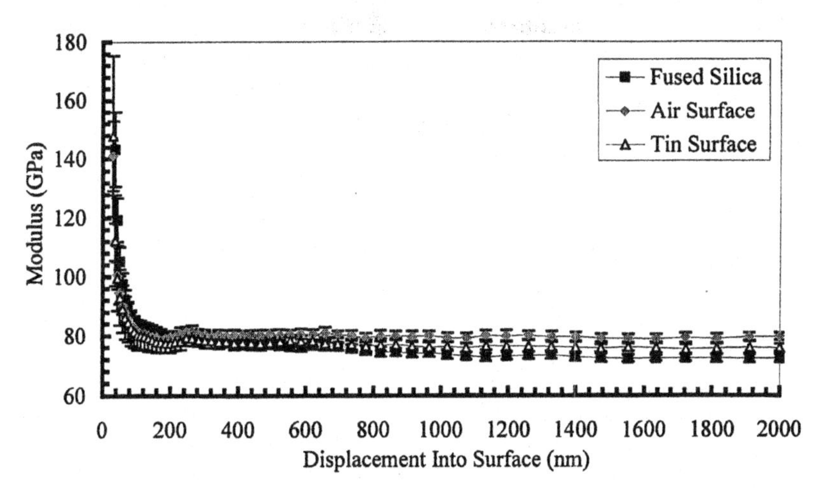

Fig. 4. Young's modulus as a function of displacement into fused silica and the air and tin surfaces of 4mm thick float glass specimens.

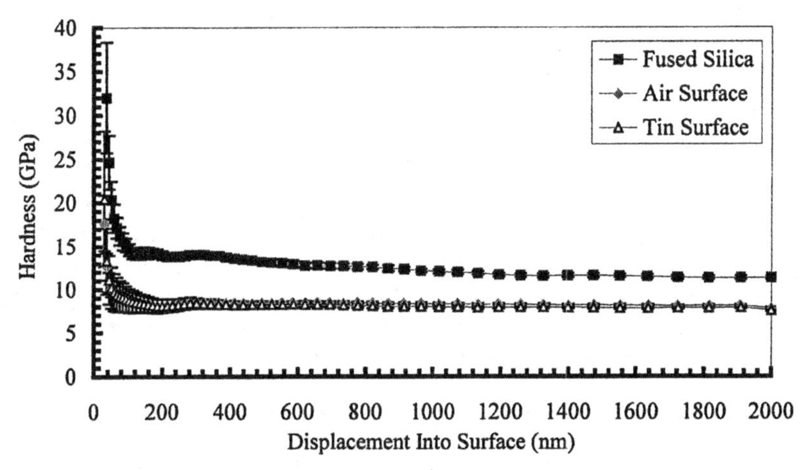

Fig. 5. Hardness as a function of displacement into fused silica and the air and tin surfaces of 4mm thick float glass specimens.

Table I. Young's modulus results for silica and the air and tin surfaces of 4mm thick float glass, calculated from twenty separate Z-contrast measurements taken either at single spot or in an array.

Sample	Frequency (GHz)	Mean v_R (ms^{-1})	θ_R (^{o})	Δz (μm)	v_{Rcorr} (ms^{-1})	E (GPa)
Spot						
Silica	1	3621	24.3	8.4		
Air surface	1	3543	24.9	8.0	3313	82
Tin surface	1	3479	25.4	7.7	3254	79
Array						
Silica	1	3662	24.0	8.6		
Air surface	1	3551	24.9	8.1	3320	82
Tin surface	1	3495	25.3	7.8	3268	79

The nanoindentation data (Figs. 4 and 5) illustrates that the variation of the first few points is noticeably higher than that of the others. This is probably due to surface imperfections; wear of the tip and slight deviations in the indenter's measurements at such shallow depths caused by vibration. In order to minimise the chance of indenting an uneven surface the sites for nanoindentation were chosen manually. All of the samples also demonstrate that for depths greater than 800nm the values for Young's modulus and hardness become more consistent and the standard deviation of the modal values reduces, this probably due to the structure and the chemical composition of the float glass becoming more uniform towards the centre of the specimen.

The Scanning Acoustic Microscope was calibrated using a sample of fused silica which had a known Young's modulus and Poisson's ratio, with distilled water at room temperature as the coupling fluid at an operating frequency of 1GHz. Twenty separate V(z) measurements were taken of this specimen all at exactly the same location (or spot) of the sample and hence an average value of v_R was determined experimentally, this was then repeated but the twenty V(z) measurements were taken in an array. From Eq. (7) Δz, the average distance was calculated, this was then compared to the theoretical value of Δz and hence a correction factor for the 1GHz lens was determined.

V(z) curves were also taken in the same way for the air and tin surfaces of specimens of 4mm thick float glass under the same operating conditions as for the calibration experiment. Assuming the wave speed for water, v_o = 1492ms^{-1}, an average Rayleigh wave velocity, v_R = 3543ms^{-1} was determined for the air surface which gave an average distance, Δz = 8.0μm from Eq. (7). Using these results in Eqs. (8 and 9), the Young's modulus for the air surface was calculated to be 82GPa. This value is marginally lower than the mean value of Young's modulus for the air surface (80GPa) that was obtained from the least variant area of Fig. 4.

The method was then repeated in order to find the value of Young's modulus for the tin surface of the glass (79GPa), and yet again a difference was found between this value and the value from the nanoindentation experiment (76GPa). A reason for this discrepancy in the results may lie in the assumption of a fixed Poisson's ratio for the material ($\nu = 0.23$) that was made for both testing methods even though Young's modulus was found to vary. An investigation by Bamber *et al.* [13], discussed the problem of requiring an accurate appraisal of Poisson's ratio for both techniques in order to calculate Young's modulus. This was resolved by comparing experimental results of Young's modulus from both procedures graphically against Poisson's ratio. A study with fused silica showed that intersection of the acoustic curve with the nanoindentation curve proved an accurate evaluation of Young's modulus and Poisson's ratio for bulk materials. However, indentations at 100nm provided inaccurate results and it was concluded that this was caused by their dependency on the area function and the model of elastic-plastic contact of a non-rigid indenter which may not be true for all depths[13].

The values in Table I. display the good agreement between the measurements taken at a single location and those taken in an array, which is true for all of the results of both the tin and air surfaces of the glass and the silica sample. However, many more "spot" and array tests would be required at different locations on the surfaces of the specimen in order to confirm the homogeneity of the sample. Potentially, the inhomogeneous area may be so large that all of the measurements could be unwittingly made within it or conversely, the inconsistency could be small enough to be missed.

The results shown in Table I. for the two surfaces of the float glass again reveal the same inconsistency between the air and tin faces observed from the nanoindentation data. The difference between the Rayleigh wave velocity v_R, of the air surface and the tin surface is comparable to the observation made by Warren *et al.* [14], where the tin side was found to have a slightly lower velocity by approximately $10ms^{-1}$ than the air surface. They concluded that this discrepancy was due to the difference in crack densities between the two sides, probably caused by the tin surface travelling along rollers during the final stages of the float process. Although the presence of diffused tin ions in the tin surface giving a denser surface layer may also lead to a reduction in Rayleigh wave velocity. These conclusions support both the nanoindentation results and the SAM results obtained in this investigation.

CONCLUSIONS

A significant difference was found between the Young's modulus values determined by nanoindentation of a 4mm thick specimen of float glass, with the tin surface producing values that were 5% lower than the air surface. This discrepancy is believed to be due to differences in near-surface composition and surface damage between the two sides. The Young's modulus of the air and tin surfaces were also determined from mean Rayleigh wave velocities of the sample, these were found to be 82GPa and 79GPa respectively regardless of whether the V(z) curves were taken from sites in an array or from a single spot, compared to the average results of 80GPa for the air surface and 76GPa for the tin surface obtained by nanoindentation. The discrepancy in these results has been caused by using an assumed Poisson's ratio for both techniques, this can be avoided in future studies if both Young's modulus and Poisson's ratio are found by combining results from nanoindentation and V(z) acoustic microscopy.

ACKNOWLEDGEMENTS

OG would like to acknowledge the support of ESPRC for a graduate studentship. We would also like to thank Paul Warren and Jonathan Williams for their advice and for providing the samples of float glass used in this work.

REFERENCES

[1]Pilkington, "http.//www.pilkington.com/".
[2]H. G. Pfaender, *Schott Guide to Glass*, Chapman & Hall, London, 1996.
[3]L. Riester, R. J. Bridge, and K. Breder, "Characterization of Vickers, Berkovich, Spherical and Cube Cornered Diamond Indenters by Nanoindentation and SFM," *Materials Research Society Symposium Proceedings,* **522** 45-50 (1998).
[4]A. E. H. Love, "The Stress Produced in a Semi-Infinite Solid by Pressure on Part of the Boundary," *Philosophical Transactions of the Royal Society of London. Series A,* **228** 377-420 (1929).
[5]A. E. H. Love, "Boussinesq's Problem for a Rigid Cone," *Quarterly Journal of Mathematics,* **10** 161-175 (1939).
[6]J. W. Harding and I. N. Sneddon, "The Elastic Stresses Produced by the Indentation of the Plane Surface of a Semi-Infinite Elastic Solid by a Rigid Punch," *Proceedings of the Cambridge Philosophical Society,* **41** 16-26 (1945).
[7]I. N. Sneddon, "The Relation between Load and Penetration in the Axisymmetric Boussinesq Problem for a Punch of Arbitrary Profile," *International Journal of Engineering Science,* **3** 47-57 (1965).
[8]W. C. Oliver and G. M. Pharr, "Improved Technique for Determining Hardness and Elastic Modulus Using Load and Displacement Sensing Indentation Experiments," *Journal of Materials Research,* **7** 1564-1580 (1992).
[9]J. C. Hay, A. Bolshakov, and G. M. Pharr, *J. Mater. Res.,* **14** 2296 (1999).
[10]J. M. Rowe, J. Kushibiki, M. G. Somekh, and G. A. D. Briggs, "Acoustic Microscopy of Surface Cracks," *Philosophical Transactions of the Royal Society (London) Series A,* **320** 201-214 (1986).

[11]J. Attal, L. Robert, G. Despaux, R. Caplain, and J. M. Saurel, Acoustical Imaging, pp. 607-616 in *International Symposium on Acoustical Imaging*, H.-P. Harjes Plenum Press, Bochum, Germany, 1991.
[12]A. Briggs, *Acoustic Microscopy*, Clarendon Press : Oxford University Press, Oxford ; New York, 1992.
[13]M. J. Bamber, K. E. Cooke, A. B. Mann, and B. Derby, "Accurate Determination of Young's Modulus and Poisson's Ratio of Thin Films by a Combination of Acoustic Microscopy and Nanoindentation," *Thin Solid Films,* **398-399** 299-305 (2001).
[14]P. D. Warren, C. Pecorari, O. V. Kolosov, S. G. Roberts, and G. A. D. Briggs, "Characterization of Surface Damage Via Surface Acoustic Waves," *Nanotechnology,* **7** 295-301 (1996).
[15]L. D. Landau and E. M. Lifshitz, *Theory of Elasticity*, Pergamon, Oxford, 1970.
[16]J. Kushibiki and N. Chubachi, "Material Characterisation by Line-Focus-Beam Acoustic Microscope," *IEEE Transactions on Sonics and Ultrasonics,* **SU-32** [2] 189-212 (1985).
[17]H. Wu, C. W. Lawrence, S. G. Roberts, and B. Derby, "The Strength of Al2o3/Sic Nanocomposites after Grinding and Annealing," *Acta materialia,* **46** [11] 3839-3848 (1998).
[18]M. Sternitzke, E. Dupas, P. Twigg, and B. Derby, "Surface Material Properties of Alumina Matrix Nanocomposites," *Acta materialia,* **45** [10] 3963-3973 (1997).
[19]P. Zinin, M. H. Manghnani, Y. C. Wang, and R. A. Livingston, "Detection of Cracks in Concrete Composites Using Acoustic Microscopy," *NDT & E International,* **33** 283-287 (2000).
[20]C. Pecorari, C. W. Lawrence, S. G. Roberts, and G. A. D. Briggs, "Quantitative Evaluation of Surface Damage in Brittle Materials by Acoustic Microscopy," *Philosophical Magazine A,* **80** [11] 2695-2708 (2000).
[21]C. Pecorari, "Modeling Variations of Rayleigh Wave Velocity Due to Distributions of One-Dimensional Surface-Breaking Cracks," *Journal of the Acoustical Society of America,* **100** [3] 1542-1550 (1996).
[22]C. Pecorari, "On the Effect of a Residual Stress Field on the Dispersion of a Rayleigh Wave Propagating on a Cracked Surface," *Journal of the Acoustical Society of America,* **103** [1] 616-617 (1998).
[23]C. Pecorari, "Rayleigh Wave Dispersion Due to a Distribution of Semi-Elliptical Surface-Breaking Cracks," *Journal of the Acoustical Society of America,* **103** [3] 1383-1387 (1998).
[24]J. C. Hay and G. M. Pharr, "Experimental Investigations of the Sneddon Solution and an Improved Solution for the Analysis of Nanoindentation Data," *Materials Research Society Symposium Proceedings,* **522** 39-44 (1998).

MICROINDENTATION OF THERMAL BARRIER COATINGS USING HIGH TEMPERATURE DISPLACEMENT SENSITIVE INDENTER

Chang-Hoon Kim and Arthur H. Heuer
Case Western Reserve University
10900 Euclid Avenue
Cleveland, OH 44106-7204

Brian D. Kernan
Massachusetts Institute of Technology
77 Massachusetts Avenue
Cambridge, MA 02139

ABSTRACT

As-deposited electron beam physical vapor deposited (EB-PVD) thermal barrier coating (TBC) samples were indented from room temperature to 900 °C using an instrumented high temperature vacuum displacement sensitive indenter (HTVDSI). The apparent hardness was obtained from the direct measurement of the size of the residual indents, and decreased from 3.2 GPa at room temperature to 0.8 GPa at 900 °C. From the load-displacement curves recorded during the indentation, the hardness as well as the elastic modulus of the TBC was calculated. The calculated hardness showed the same temperature dependence as that measured optically. The columnar microstructure of the TBCs led to lower hardness and lower elastic modulus than bulk material of the same composition.

INTRODUCTION

Thermal barrier coatings (TBCs) are currently used to provide thermal insulation to superalloy components of gas turbine engines in high temperature environments, thereby improving the engine efficiency and performance [1]. As usage of the gas turbine accelerates, premature failure of TBCs, usually resulting from spallation or erosion damage, occurs. Understanding the failure mechanisms of TBCs is crucial to improving their reliability.

The hardness and elastic moduli of TBCs have been investigated to study plastic and elastic deformation. [2-5]. It is known that TBCs show lower values in both their hardness and elastic modulus than dense ceramics or single crystals [6,7], due to their porous microstructure. In a conventional microhardness measurement, the hardness is obtained by direct imaging of the residual indents. However, large errors could be introduced in the measurement when the indent is small [8]. In addition, it is difficult to measure the size of indents when they have an irregular shape that might result from the porous microstructure of the sample [9]. Finally, unusual relaxation behavior on unloading might result from the unusual microstructure of TBCs.

The displacement sensitive indentation technique, which has been given much attention recently, monitors the load and displacement experienced by the indenter as it is driven into and withdrawn from the sample. From the recorded load-displacement curve, the hardness and elastic modulus can be determined; of course, the modulus cannot be determined by conventional (i.e. lacking load-depth data) hardness measurements [10-12].

In the present study, an instrumented high temperature vacuum displacement sensitive indentation (HTVDSI) equipment was used to indent the yttria-stabilized zirconia (YSZ) top coat in an as-deposited TBC from room temperature to 900 °C. In addition to the hardness obtained by the conventional method, the hardness and elastic modulus of TBC were calculated from the load-displacement curves.

EXPERIMENTAL PROCEDURE

The as-deposited TBC samples consisted of a 3 mm thick nickel-based superalloy substrate, onto which a 50 μm thick PtNi aluminide bond coat layer had been formed; this was followed by electron beam physical vapor deposition (EB-PVD) of a 125 μm thick ZrO_2-7wt% Y_2O_3 (YSZ) top coat. The well-known columnar microstructure of the EB-PVD YSZ top coat is shown in Fig. 1; the column size is about 5~10 μm at the column tops. For the microindentation experiment, the top surface of the TBC was polished, finishing with 1 micron diamond paste. Approximately 10 μm of material was removed from the top coat during polishing.

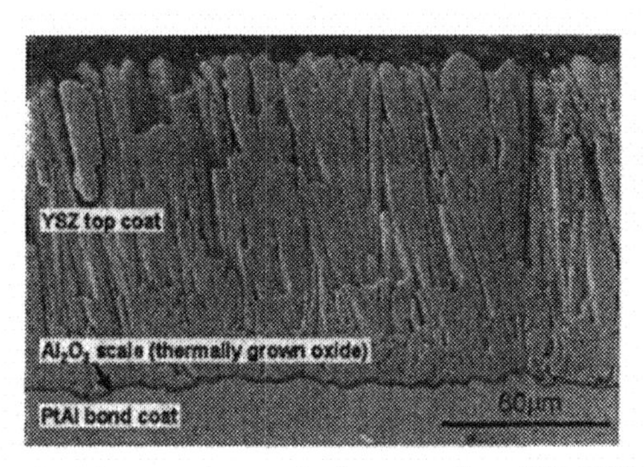

Fig. 1. Scanning electron micrograph of cross section of EB-PVD TBC.

A schematic diagram of the indenting mechanism in our HTVDSI equipment is shown in Fig. 2. The indenter is a conventional Vickers diamond pyramid. Rather than applying a dead load, a current-controlled electromagnetic moving coil is used to apply the load. This enables loading at a wide range of rates and reasonable maximum loads. The position of the indenter is determined by capacitance displacement gauges, which can detect displacement changes of 25 nm. The load resolution is about 0.25 g ($\approx$2.5 mN). The sample and the indenter are heated separately to keep both temperatures the same, and the whole system is enclosed in a vacuum chamber. The entire heating/loading/unloading cycle, including data recording, is computer controlled.

The as-deposited TBC samples were indented in the temperature range 25~900 °C at a maximum load of 100 g ($\approx$1 N) with a 2.5 s load duration time.

The displacement rate of the indenter was about 1.5~2 μm/s.

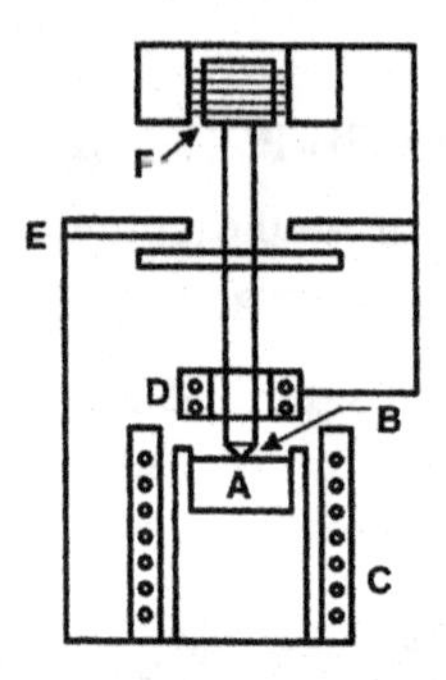

Fig. 2. A schematic diagram of the indenting mechanism of this HTVDSI: (A) sample, (B) indenter, (C) sample furnace, (D) indenter furnace, (E) capacitance displacement gauge, (F) load application coil.

RESULTS AND DISCUSSION

Fig. 3 shows typical load-displacement curves recorded at 25 °C and 600 °C. As the temperature is increased, the indenter penetrates further into the sample, and plastic deformation occurs via indentation creep if the load is held constant. On unloading, there occurs some elastic recovery in the indent.

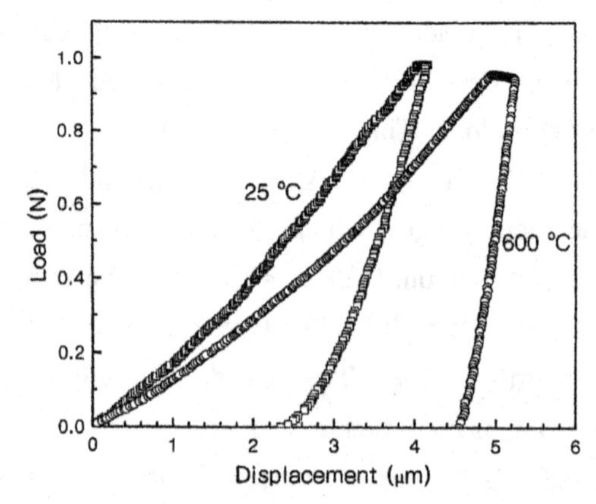

Fig. 3. Load-displacement curves of the as-deposited TBC for a maximum load of 100 g (≒1 N).

While the hardness is usually obtained by the direct measurement of the diagonals of the residual indents, it can also be calculated from the load-displacement curve [10]. The hardness is defined as

$$H = \frac{P_{max}}{A} \tag{1}$$

where P_{max} is the maximum load and A is the contact area of indentation. Here, the contact area can be either a projected or a surface area; in fact, the definition of Vickers hardness uses the surface contact area. While the Vickers hardness is calculated with the contact area measured on unloading, i.e., after some recovery has taken place at the indent, the contact area in Eq. (1) refers to that in the fully loaded condition. This contact area is calculated from the contact depth, h_c, which is defined as

$$h_c = h_{max} - \varepsilon \frac{P_{max}}{dP/dh} \tag{2}$$

where h_{max} is the maximum displacement (or depth) determined experimentally, ε is a geometric constant (ε=0.72 for a conical indenter), and dP/dh is the initial slope of the unloading curve.

The hardness of TBCs calculated using this analysis, as well as that measured optically, is shown in Fig. 4. The optically measured hardness decreases from 3.2 GPa at 25 °C to 0.8 GPa at 900 °C, considerably lower than that of dense ZrO_2 ceramics or single crystals (12~14 GPa) [6,7]. The considerable compliance resulting from the extensive porosity and separation between the isolated columns shown in Fig. 1 yields a relatively low hardness. Also, the hardness measured in HTVDSI is slightly higher than that measured in a commercial hot hardness tester [2], which is attributed to the much lower loading rate in the HTVDSI; the loading rate in the commercial tester was 200 μm/s.

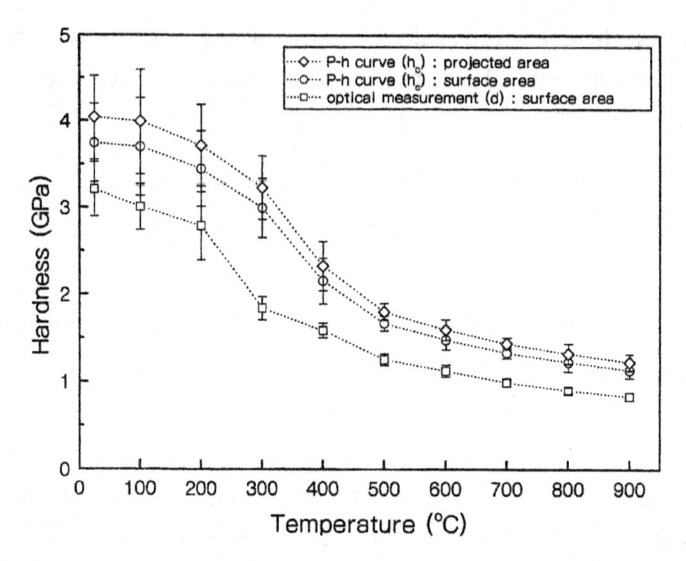

Fig. 4. Hardness vs. temperature plots of the as-deposited TBCs. Squares are obtained by optical measurement of the indents, and circles and diamonds are calculated from the load-displacement curves.

The calculated hardness is higher than that measured optically, but displays the same temperature dependence. The discrepancy of hardness between those two methods is attributed to the elastic recovery of the indent during unloading [7,10].

In addition to the hardness, the elastic modulus can also be calculated from the load-displacement curve. The initial slope of the unloading curve, *dP/dh*, is related to a reduced modulus, E_r, through the equations

$$\frac{dP}{dh} = \frac{2}{\sqrt{\pi}} E_r \sqrt{A} \tag{3}$$

$$\frac{1}{E_r} = \frac{(1-\nu^2)}{E} + \frac{(1-\nu_i^2)}{E_i} \tag{4}$$

where E (E_i) and $\nu(\nu_i)$ are elastic modulus and Poisson's ratio for the sample (indenter). The calculated modulus of TBC is shown in Fig. 5; there is, as expected, little change over this temperature range. The modulus is much lower than the elastic modulus of YSZ ceramics (190~210 GPa), which we are certain is not an instrumental artifact. The apparent lower elastic modulus of TBCs is also attributed to the porous and columnar microstructure [4].

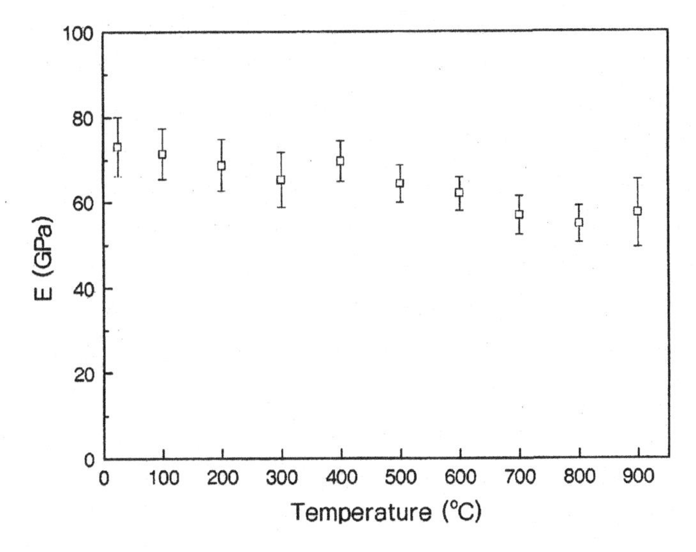

Fig. 5. Elastic modulus vs. temperature plot of the as-deposited TBCs.

CONCLUSIONS

An instrumented high temperature vacuum displacement sensitive indenter was used for microindentation on YSZ top coats on as-deposited TBCs. The displacement sensitive indentation technique enables the determination of both hardness and elastic modulus. TBCs are softer and more compliant than single crystal or polycrystalline ZrO_2 of similar composition, which is a result of their porous columnar microstructure.

ACKNOWLEDGEMENTS

The authors acknowledge the Office of Naval Research for their financial support.

REFERENCES

[1]N.P. Padture, M. Gell, and E.H. Jordan, "Thermal Barrier Coatings for Gas-Turbine Engine Applications," *Science*, **296** 280-84 (2002).

[2]B.D. Kernan, A. He, and A.H. Heuer, "Microstructural Evolution and Microhardness in Zirconia-Based EB-PVD Thermal Barrier Coatings," *submitted to J. Am. Ceram. Soc.*

[3]C.A. Johnson, J.A. Ruud, R. Bruce, and D. Wortman, "Relationships Between Residual Stress, Microstructure and Mechanical Properties of Electron Beam-Physical Vapor Deposition Thermal Barrier Coatings," *Surface and Coatings Technology*, **108-109** 80-85 (1998).

[4]J.P. Singh, M. Sutaris, and M. Ferber, "Use of Indentation Technique to Measure Elastic Modulus of Plasma-Sprayed Zirconia Thermal Barrier Coating," *Ceramic Engineering and Science Proceedings*, **18** [4] 191-200 (1997).

[5]J.I. Eldridge, D. Zhu, and R.A. Miller, "Mesoscopic Nonlinear Elastic Modulus of Thermal Barrier Coatings Determined by Cylindrical Punch Indentation," *J. Am. Ceram. Soc.*, **84** [11] 2737-39 (2001).

[6]G.N. Morscher, P. Pirouz, and A.H. Heuer, "Temperature Dependence of Hardness in Yttria-Stabilized Zirconia Single Crystals," *J. Am. Ceram. Soc.*, **74** [3] 491-500 (1991).

[7]J. Alcala, "Instrumented Micro-Indentation of Zirconia Ceramics," *J. Am. Ceram. Soc.*, **83** [8] 1977-84 (2000).

[8]J. Ricote, L. Pardo, and B. Jiménez, "Mechanical Characterization of Calcium-Modified Lead Titanate Ceramics by Indentation Methods," *J. Mater. Sci.*, **29** 3248-54 (1994).

[9]L.B. Harris and F.K. Nyang, "Porosity, Density and Hardness of Y-Ba-Cu High-T_c Superconductors," *J. Mater. Sci. Lett.*, **7** 801-03 (1988).

[10]W.C. Oliver and G.M. Pharr, "An Improved Technique for Determining

Hardness and Elastic Modulus Using Load and Displacement Sensing Indentation Experiments," *J. Mater. Res.*, **7** [6] 1564-83 (1992).

[11]G.M. Pharr and R.F. Cook, "Instrumentation of a Conventional Hardness Tester for Load-Displacement Measurement During Indentation," *J. Mater. Res.*, **5** [4] 847-51 (1990).

[12]M.F. Doerner and W.D. Nix, "A Method for Interpreting the Data from Depth-Sensing Indentation Instruments," *J. Mater. Res.*, **1** [4] 601-09 (1986).

RESIDUAL STRESS BEND EFFECT DUE TO DIAMOND-TIP SCRIBING OF AN Al_2O_3-TiC COMPOSITE CERAMIC

B. W. Austin
IBM Corporation
Fishkill NY 12524

T. Randall and R. O. Scattergood
Materials Science and Engineering Dept.
North Carolina State University
Raleigh, NC 27695-7907

ABSTRACT

A system consisting of a microhardness tester and a motorized stage was used to produce controlled diamond scribes on Al_2O_3-TiC ceramic samples. A bending distortion is produced as a result of the residual stresses generated by the scribing operation. The change in sample shape was measured using an optical profilometer. Data from the measurement tool were analyzed using computer software to extract the net bending effect. An analytical model for the bend effect based on the residual stresses due to line-force dipoles was developed. The adjustable parameter in the model, dipole strength, was determined using bend-angle measurements. The predicted bend-displacement profiles were compared with measured profiles and the agreement was excellent. Although fracture could be observed along the flanks of scribe traces, this did not appear to diminish the residual stress intensity produced over the range of scribe loads used. The line-force dipole model has application for process analysis or as a benchmark for the verification of detailed elastic-plastic models of scribe-induced deformation.

INTRODUCTION

Indentation methods for the study of fracture processes in ceramic materials have been the subject of many investigations. Comprehensive reviews are available [1-3]. Scribing using sharp diamond tips is related to single indentations in that the loading conditions are spatially localized. The stress fields produced by indentation or scribing can be distinguished as those present during the loading-unloading cycle and those remaining after the loads are removed, i.e., the residual stress fields. Both types of stresses can contribute to the fracture processes that result. Various models have been developed to describe indentation-type stresses.

In the case of single indentations, the elastic-plastic analysis of Chiang et. al. [4] stands as a pioneering study of the detailed deformation mechanisms that control both the loading-unloading stresses and the residual stresses. Phenomenological models based on single or dipole point forces have also been developed. Among these, the model developed by Yoffe [5] stands as a pioneering study for the loading-unloading and residual stresses produced for single indentations. Force dipole models can be quite useful because they produce an approximate, closed-form solution for the stress fields. However, a direct measurement of the parameters that dictate the scale of the stresses or stress-intensity factors in force-dipole models is difficult. In the work reported here, a force-dipole model for the residual stresses resulting from scribing is developed. By making scribes on plate samples, and measuring the bend deflections resulting from the bending moment produced by the residual stresses, a dipole-strength parameter can be determined experimentally. This provides a unique way to calibrate the force dipole model. The procedure can be used to investigate the effects of loading conditions and material response on scribing, as evidenced by concomitant changes in the dipole strength parameter. This approach can also provide benchmarks for the verification of more detailed elastic-plastic models for scribing. In the following, the procedure for measuring the residual stress bend-effect is presented along with the line-force dipole model. The focus material for the experimental work was an Al_2O_3-TiC ceramic composite. For comparison, some preliminary data on silicon is also presented.

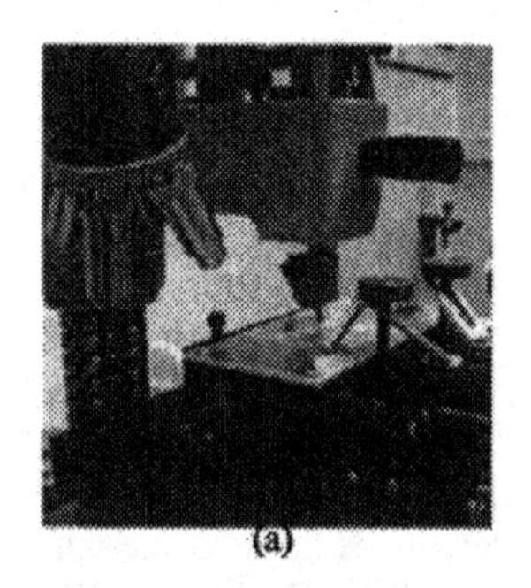

(a)

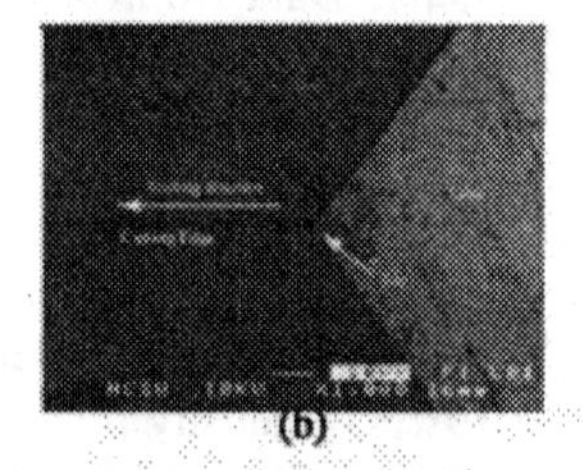

(b)

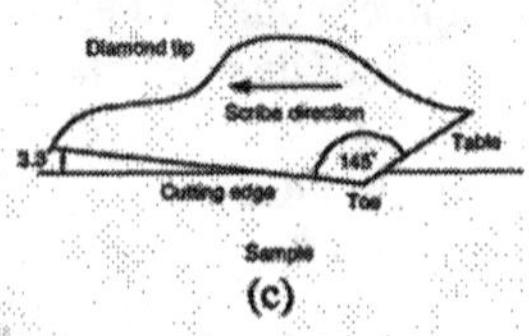

(c)

Figure 1. (a) Scribing setup. (b) Dynatex V4-64 diamond scribing tool. The angle between the faces defining the cutting edge is 90°. (c) Schematic of the scribe-tip geometry.

EXPERIMENTAL PROCEDURE

A Zwik microhardness tester shown in Fig. 1a was used as the platform for the scribing tests. To produce a scribe at a constant speed, a motorized translation stage was fixed to the base of the hardness tester. The system was fixtured so that the direction of motion of the stage could be aligned with the cutting direction for a Dynatex V4-64 diamond scribing tool shown in Fig. 1b. This is a three-faced pyramidal diamond tool used in the semiconductor industry for scribing. For the tests done here, the Dynatex tool shank was mounted such the cutting edge made an angle of approximately 3.3° from the Al_2O_3-TiC (AlTiC) material surface, as is shown in Fig. 1c. The normal force on the scribing tip was controlled by placing weights on the loading pan of the microhardness tester. The normal force W ranged from 0.3 to 1.7 N.

Higher loads caused excessive wear and chipping damage on the scribing tip. SEM observations of the tip edge were made during the tests and the tip was replaced whenever damage or excessive wear was observed.

The AlTiC ceramic used for the tests contained 65 vol% Al_2O_3 and 35vol% TiC. The microstructure consists of polycrystalline Al_2O_3 with a grain size of about 1 μm and TiC particles about 1 μm in diameter. The samples used for the scribing tests were polished AlTiC rectangular plates measuring approximately 1.0 mm x 1.0 mm x 0.3 mm. The spacing between the scribes was approximately 0.2 mm. For standard tests, three scribes were made across the width of a sample on the polished side with the middle scribe centered on the sample span. A Zygo New View non-contacting optical profilometer was used to measure the 2D surface height profiles of the test samples. Measurements were taken before and after scribing. To capture the intrinsic bend effect, profilometer data taken before scribing must be subtracted from the data taken after scribing. This was accomplished by importing both sets of data into a file and aligning fiducial marks. The initial data were subtracted from the final data, point by point, using the JMP software package. Any remaining tilt was removed from the data by the software. The sample surfaces on the measurement side need to have an optical quality surface finish. Details are reported in [6]. Fig. 2 shows a typical example of the 2D profilometer data obtained using three scribes per sample. The scribe traces were made on the top surface of the samples and the bending occurs downward. The steep craters present in Fig 2 occur at the three scribe positions. These are due to drop-out of the optical profilometer signal at the high-slope regions adjacent to scribe traces. As a final step, 1D profiles were extracted from the 2D data in order to obtain a simple representation of the bend-displacement profiles orthogonal to the scribe direction.

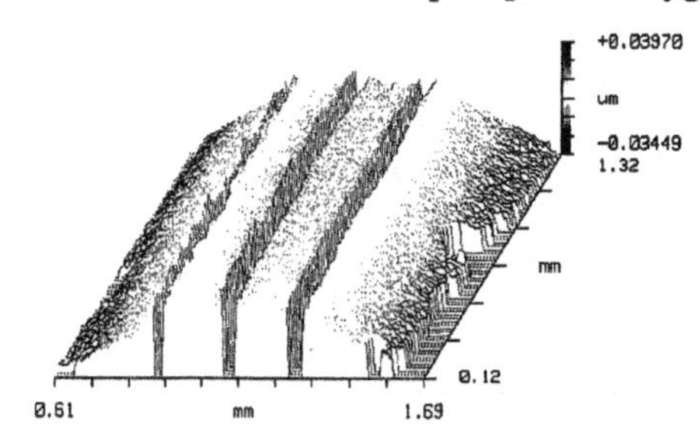

Figure 2. Optical profilometer data for a scribed sample.

BEND-EFFECT MODEL

The action of the residual stresses produced along a scribe trace using a scribe (normal) load W can be modeled as the line-force dipole shown in Figure 3a. The dipole forces act outward from the scribe trace because the elastic unloading of the plastically deformed and compressed material produces outward wedging forces on the surrounding restraining elastic material. A pair of orthogonal dipole forces acting on the surface of an elastic half space was introduced by Yoffe [5] as a model for the residual stresses due to single indentations. Ahn, et. al. [7] analyzed the stresses due to scribing as the superposition of a row of overlapping

Yoffe-type indentation dipole force pairs. For scribing across the width of a plate sample, superposition of the dipole force pairs is mathematically equivalent to a line force dipole because the dipole forces oriented along the trace mutually cancel, as is indicated in Figure 3b. Line-force dipole stresses for an isotropic elastic half space can be readily obtained from standard elasticity solutions. The dipole forces in Fig. 3 will produce a bending moment around the scribe direction in a finite-size plate sample. The bend deflection and bend angle along the plate length, $0 \leq x \leq L$, are denoted as $\delta(x)$ and $\phi(x)$, respectively.

For the analysis of the bend effect in finite-size plates, imagine that the 2L x h x b plate shown in Fig. 3 is extracted from the elastic half-space. The finite-body stress solution requires that the stress-field tractions due to the line-force dipole vanish on the plate free surfaces. If the plate thickness h is reasonably small compared to its length 2L, the stresses acting on the plate edges are negligible and their effects can be ignored. However, the stresses acting on the bottom surface of the plate cannot be ignored. Since there is no resultant force due to dipoles, the equilibrium configuration can be approximated by applying reversed stresses to the bottom of the plate and then solving a beam-bending problem using these stresses as loading functions. This procedure will insure equilibrium of forces and moments acting on any cross section of the plate. For internal stress fields, this has been shown to be a very good approximation to the exact finite-body solution [8]. It is important to include both the normal and shear stresses for the beam loading functions since these have comparable magnitudes for the bending effect. The detailed analysis is given in the Appendix. FEM calculations were made which showed that the bend effect predicted by this approach is within a few percent of an exact, finite-body numerical solution [6]. The beam-bending solutions for a single scribe centered on the plate in Fig. 3 are given by eq. [1] (also see eq. [A6]). E is the elastic modulus (E = 405 GPa for AlTiC), h is the plate thickness, B = Fa is the line-force dipole strength parameter and X = x/h is the normalized distance along the plate. B is the adjustable parameter in the model and is represented here in the infinitesimal dipole limit, i.e., B = Fa = constant in the limit where a -> 0 as F -> ∞. Since the problem can be treated as simple (plane strain) bending of a rectangular beam, the results are independent of the plate width b.

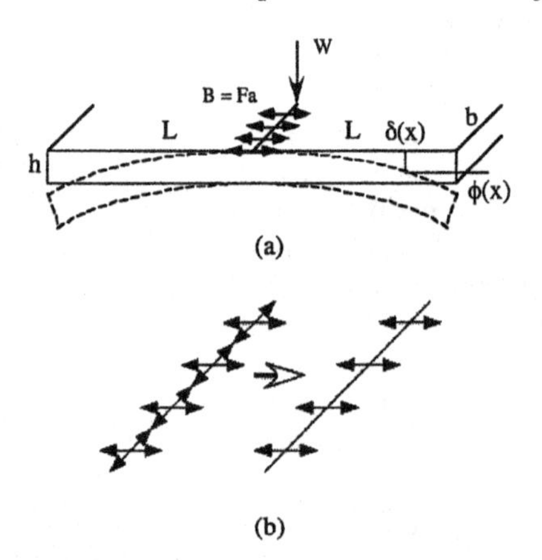

Figure 3. (a) Line-force dipole and the parameters used to describe the bend effect. (b) Superposition of a row of Yoffe-type orthogonal (crossed) dipole force pairs produces a line force dipole.

$$\phi(X) = -\frac{6B}{\pi Eh^2}\left[\tan^{-1}(X) + \frac{X}{1+X^2}\right] = \frac{6B}{\pi Eh^2}f(X)$$
$$\delta(X) = -\frac{6B}{\pi Eh}\left[X\tan^{-1}(X)\right] = \frac{6B}{\pi Eh}g(X) \tag{1}$$

The bend-effect equations for multiple scribe traces are easily obtained from the superposition of the single-scribe functions in eq. (1), suitably offset in the X coordinate to account for the change of scribe position. The scribing geometry considered in the Appendix is a central scribe surrounded by specified number of equally spaced scribes on either side of the central one. The normalized bend angle f(X) and bend displacement g(X) functions are plotted in Fig. A4 for single and multiple scribes. Multiple scribes increase the overall deflection due to the bend effect and thereby reduce scatter in the measured profilometer data. Three scribes per sample were used as the standard test in this study, and this gave very reproducible data.

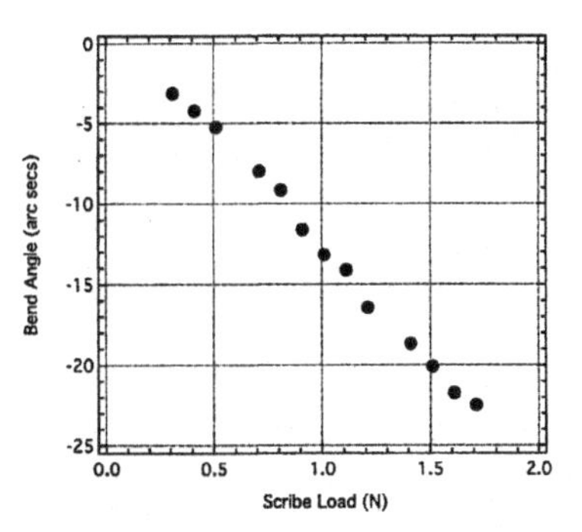

Figure 4. Bend angle ϕ vs. scribe load W data obtained for AlTiC.

The key parameters for scribing tests are the material, the scribe-tip geometry and the normal load W applied during scribing. Scribing speed was found to be a second order effect over the range of speeds tested here, 50 to 250 μm/s, and the testing was done at a nominal speed of 100 μm/s. Values of B as a function of W are a direct measure of the intensity of the residual stresses produced by a scribe load W for given scribing conditions.
To judge the overall effectiveness of the line-force dipole model, the measured bend profiles are compared to the model predictions.

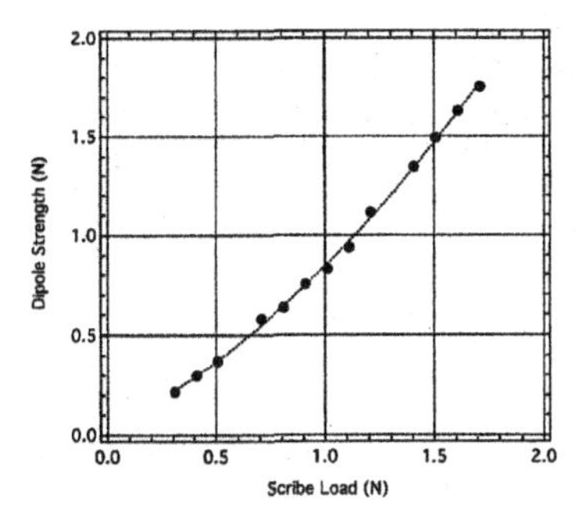

Figure 5. Dipole strength B vs. scribe load W for AlTiC.

RESULTS AND DISCUSSION

The dipole strength parameter B was determined at each scribe load by comparing measured data for the average bend angle ϕ per scribe with model predictions using the procedure described in the Appendix. The ϕ vs W test data obtained for scribing AlTiC are shown in Fig. 4. The reproducibility of the results was very good, as long as the diamond tips remained undamaged. The B vs W curve shown in Fig. 5 was obtained from the bend-angle data in Fig. 4. The curve fit through the data points in Fig. 5 is a second-order polynomial, B

$= 0.0332 + 0.516W + 0.291\ W^2$. This can be viewed as the "calibration curve" for the line-force dipole model for given diamond tool geometry, scribing conditions and material. The effectiveness of the model can be demonstrated by comparing the measured bend profiles with predicted profiles derived using the calibration curve. Fig. 6 a-c shows typical examples of measured (solid) and predicted (dashed) bend profiles for low, medium and high values of the scribe load W. The large spikes on the measured profiles are due to the profilometer drop outs at the scribe positions.

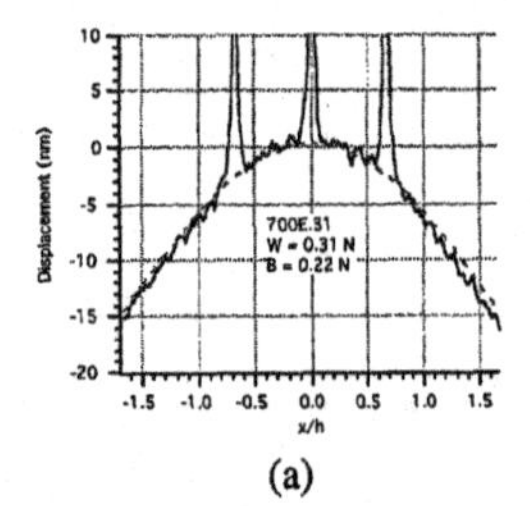

(a)

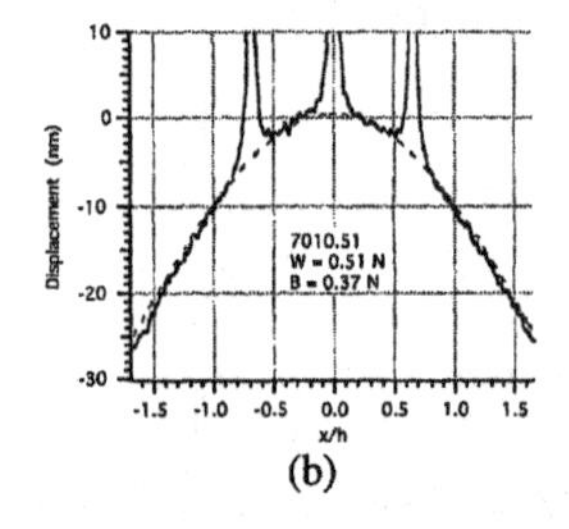

(b)

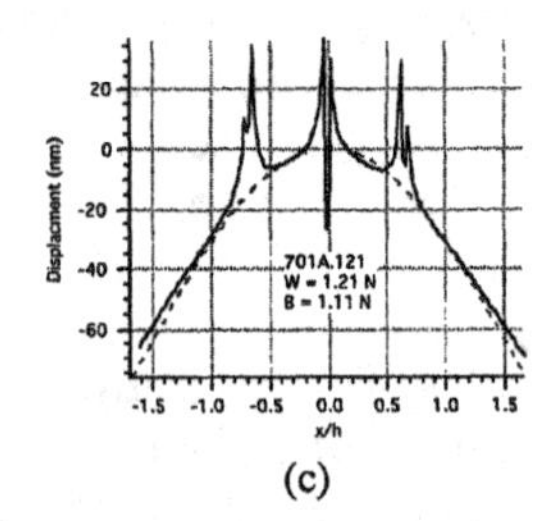

(c)

Figure 6. Measured (solid) and predicted (dashed) bend profiles for the following test conditions: (a) W = 0.31 N, p = 0.2 mm, 2L = 1.1 mm and h = 0.3 mm. (b) W = 0.51 N, p = 0.2 mm, 2L = 1.2 mm and h = 0.3 mm. (c) W = 1.21 N, p = 0.2 mm, 2L = 1.0 mm and h = 0.3 mm.

The bend profiles have the appearance of a plastic hinge around the region of the scribe traces. The measured profiles are matched to a high degree of accuracy by the model. This was the case for the profiles obtained over the entire range of scribe loads tested. As further comparison, the bend profiles shown in Fig. 7 a-b were also obtained. These were done using test conditions that were different from the standard three-scribe tests used to obtain the calibration curve in Fig. 5 (different sample dimensions and five scribes per sample). Excellent agreement is again obtained. Note that the magnitude of the bend deflections is in the nanometer range for the scribing conditions and sample sizes used.

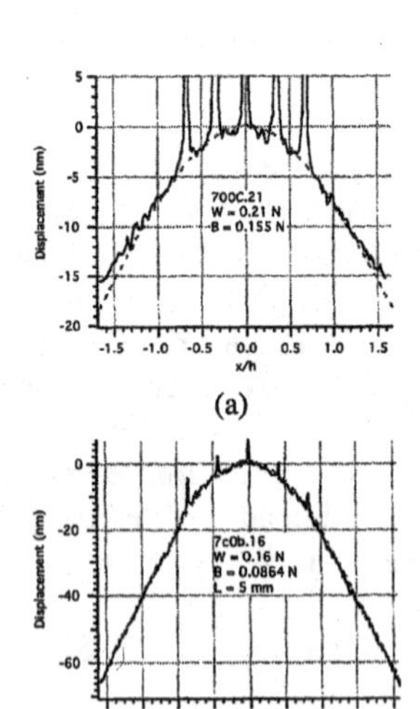

Figure 7. Measured (solid) and predicted (dashed) bend profiles for the following test conditions: (a) W = 0.21 N, p = 0.1 mm, 2L = 1.0 mm and h = 0.3 mm. (b) W = 0.16 N, p = 0.5 mm, 2L = 5.2 mm and h = 0.3 mm.

Fig. 8 a-c shows SEM micrographs of the scribe traces for samples corresponding to the test conditions and displacement profiles shown in Fig. 6. With increasing scribe load W, the scribe widths become larger and there is an increasing amount of fracture damage evident along the flanks of the scribe traces. However, because of the relatively high toughness of the AlTiC ceramic samples ($K_{IC} = 4.2$ MPa-m$^{1/2}$), extensive chipping due to lateral cracking is not evident. The central regions of the scribes still retain significant amounts of ductile-like deformation. This is most clearly evidenced in the dipole strength results shown in Fig. 5. There is no break or leveling-off in the dipole-strength curve that would indicate, for example, the onset of a severe fracture/chipping process. The latter would remove large amounts of the plastically deformed material from the scribe zone, thereby reducing the capacity to store residual stresses. One might expect this kind of change to occur at a sufficiently large value of the scribe load. However, loads beyond 1.8 – 2 N could not be used for the scribing tests on AlTiC because of catastrophic damage to the scribe tips. During the initial phases of the testing, a noticeable leveling-off did appear for the B vs. W curve in Fig. 5 when the scribe load was in the range of 2 N or higher, but it was determined that this effect results from scribing with a badly fractured diamond tip. The blunt nature of the fractured scribe tip significantly affects the deformation processes and this was reflected in a reduced dipole strength.

The line-force dipole model has several important ramifications for the study of residual stresses produced by scribing. The dipole strength parameter B gives a simple characterization of the scale of the residual stress as a function of the loading conditions, tool geometry and materials. This is a scaling parameter, in the spirit of Yoffe's original suggestion for the use of force dipoles to model indentation stress fields [1, 5]. Scribing processes are more complex and difficult to model than indentation processes. The approach presented here provides an experimental means to characterize the residual stresses produced by scribing in terms of a force-dipole strength parameter B. This should be a useful benchmark to compare with more detailed elastic-plastic models of scribing. Such comparisons were not within the scope of this study.

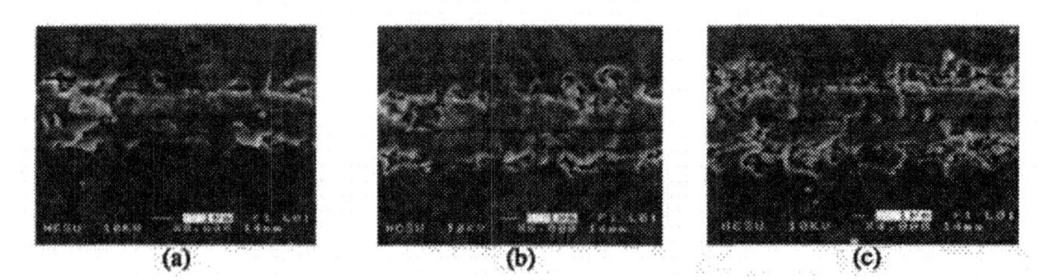

Figure 8. SEM micrographs of the scribe traces corresponding to conditions a-c in Fig. 6, respectively.

A noteworthy feature of the line-force dipole model was observed in our recent, preliminary work on the scribing of silicon wafer samples [9]. Evidently, because of the low toughness of silicon compared to AlTiC, the B vs. W curve for silicon in Fig. 9 shows a clear break-point at the onset of fracture (arrow). The latter was established visually using optical microscopy and AFM. The break-point corresponds to a critical load for the onset of a ductile-brittle transition during scribing of silicon. Based on these preliminary results, one can anticipate that the force-dipole bend effect model could be a useful technique to determine critical loads for fracture transitions in brittle materials as a function of the tool geometry and scribing conditions. In turn, this would provide critical load or critical depth parameters relevant for ductile-regime machining and precision grinding of ceramics [10-12]. Although not attempted in this study, there is also the possibility for developing a bend-effect model appropriate for linear rows of widely spaced (single) indentations in order to determine dipole strength parameters for residual stresses due to these types of indentation processes.

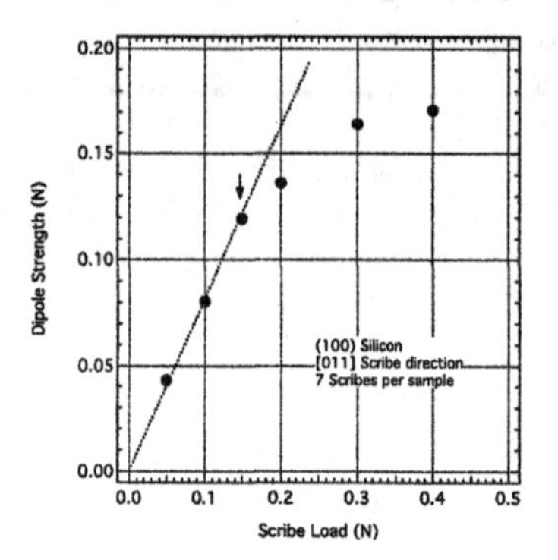

Figure 9. Dipole strength B vs. scribe load W curve for silicon for the same setup, Dynatex scribing tool and scribing geometry used for AlTiC (Fig. 1). 2L = 1 cm, h = 0.5 mm, a = 0.1 mm and seven scribes per sample. (100) crystal surface normal, [011] scribing direction. $E = E_{[100]} = 150$ GPa. The arrow indicates the onset of fracture as observed by optical and AFM microscopy.

Finally, it should mentioned that the bend effect for AlTiC has a unique application. This material is used for the manufacture of hard-disk read-write heads. Among other methods, diamond-tip scribing has been used to produce nano-scale curvature adjustments during the manufacture of the heads, which in current generation have dimensions close to the sample dimensions, 1mm x 1mm x 0.3 mm, used for the work reported here. During the manufacturing process, scribes are made on a non-functional region of the read-write head in order to tailor the curvatures for optimization of the aerodynamic (flying-height) performance [13]. A line-force dipole model can serve as the basis of a process model for predicting the bending deflections for controlled curvature adjustments during manufacturing [14].

CONCLUSIONS

A line-force dipole model gives an excellent representation of the bend-effect produced in rectangular plates by the scribe-induced residual stresses. Characterization of these stresses in terms of a dipole strength parameter will be useful for analyzing the scribing process and as a benchmark for comparing to more detailed elastic-plastic models. Based on preliminary results for silicon, it also appears that trends in the dipole strength parameter as a function of scribe load can be used as an indicator of ductile-brittle transitions for specified scribing conditions.

APPENDIX

Consider a line-force dipole acting on the surface of an elastic half-space shown in Fig. A1. The line-force dipole lies along the z-axis. The forces act outward because the plastically displaced subsurface material formed will, upon unloading, push outward on the surrounding elastic material. This is the action that produces a bend effect for diamond-tip scribing. Note that other types of "scribing processes", for example focused-laser melting, could reverse the sign of these residual-stress forces [13,15].

The stresses needed to describe the bending effect for plate scribing are σ_{yy} and σ_{xy}. The stresses due to a single line force of magnitude F (N/m), acting at the origin in the positive x direction, are given by Johnson [15] as follows

$$\sigma_{yy}(x,y) = -\frac{2F}{\pi}\frac{xy^2}{\left(x^2+y^2\right)^2}$$

$$\sigma_{xy}(x,y) = -\frac{2F}{\pi}\frac{x^2y}{\left(x^2+y^2\right)^2} \qquad \text{(A1)}$$

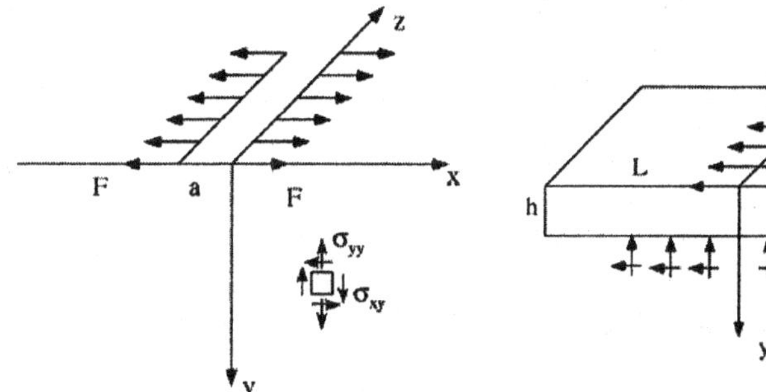

Figure A1. A single line force per unit length acting at the origin on the surface of an elastic half space produces a 2D stress field σ_{ij}. The line-force dipole field is obtained by the superposition of the two oppositely directed line forces shown where a is the spacing.

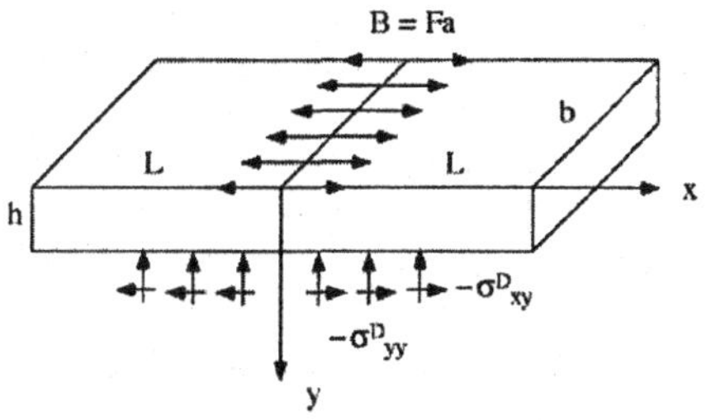

Figure A2. The finite plate is "cut out" from the elastic half space. The reversed dipole half-space stresses are applied to the bottom surface as indicated. Equilibrium of moments is obtained by allowing the plate to undergo bending with these stresses acting as the loading functions.

In order obtain the line-force dipole stress field, consider an oppositely directed line force F acting at x = -a in Fig. A1. The dipole stresses are the superposition of the stresses due to the two line forces. The infinitesimal line-force dipole stresses are obtained by taking the limit a -> 0 and F -> ∞ such that B = Fa remains finite. B is defined as the dipole strength (N). The infinitesimal line-force dipole stresses, σ^D_{yy} and σ^D_{xy}, can be obtained from the derivatives of eq. [A1] as follows

$$\sigma^D_{yy} = -a\frac{\partial\sigma_{yy}}{\partial x} = \frac{2B}{\pi}\frac{y^2\left(-3x^2+y^2\right)}{\left(x^2+y^2\right)^3}$$

$$\sigma^D_{xy} = -a\frac{\partial\sigma_{xy}}{\partial x} = \frac{4B}{\pi}\frac{xy\left(y^2-x^2\right)}{\left(x^2+y^2\right)^3} \qquad \text{(A2)}$$

For the analysis of the bending effect due to scribe-induced residual stresses, consider a finite plate with a scribe along its centerline (z axis) "cut out" from the half-space as shown in Fig. A2. The finite-body stress solution requires that the stress field due to the line-force dipole vanish on the plate surfaces. If the plate thickness h is relatively small compared to its length 2L and width b, the magnitude of the half-space stresses, σ^D_{yy} and σ^D_{xy}, acting on the plate edges is negligible and their effects can be ignored. However, the stresses (tractions) acting on the bottom surface of the plate cannot be ignored. Since there is no resultant force due to the line-force dipole, the equilibrium configuration can be approximated by applying the reversed stresses, $-\sigma^D_{yy}$ and $-\sigma^D_{xy}$, to the bottom of the plate and then solving a simple (plane strain) beam-bending problem with these stresses as the loading functions. However, this will not produce an exact solution for the stress distribution within the finite plate because there remains a self-equilibrated distribution of tractions along the free surfaces. Nevertheless, FEM calculations showed that the bending effect calculated by this procedure was within a few percent of the exact solution [6].

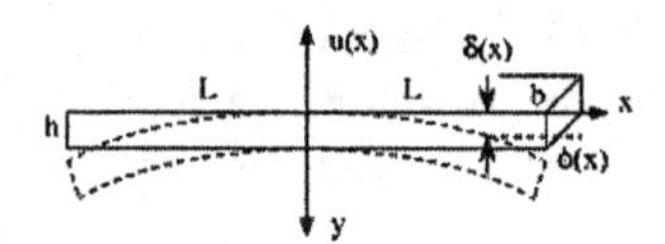

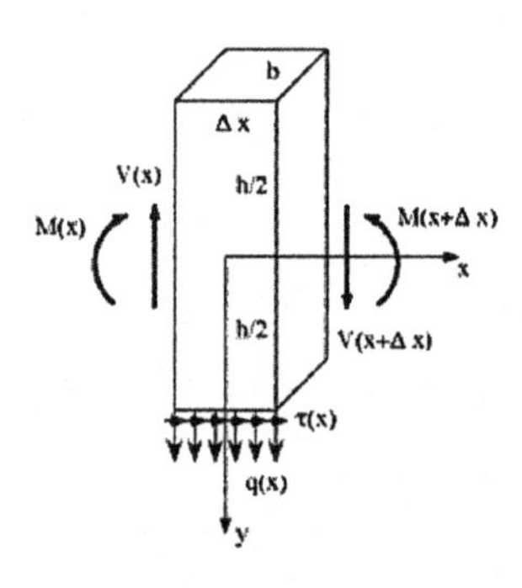

Figure A3. Top) Geometry used to describe the plate bending effect. Bottom) Shear forces and moments on a section Δx due to the normal and shear loading functions q(x) and τ(x).

The required beam-bending equations are based on Fig. A3. It is important to note that both the normal and shear tractions must be taken into account when calculating the slope and deflection of the plate. For a single, central scribe across the width b of the plate in Fig. A2, the beam equations are obtained from the equilibrium of the shear force V(x) and moment M(x) acting on an element Δx of the plate

$$\frac{dV}{dx} = -q(x)$$
$$\frac{dM}{dx} = V(x) - \tau(x)\frac{h}{2} \qquad \text{(A3)}$$

q(x) and τ(x) are the normal and tangential beam loading functions, respectively. These are obtained from the reversed half-space stresses given in eq. [A2] using the value y = h

$$q(x) = -b\sigma_{yy}^{D}(x,h) = -\frac{2Bb}{\pi}\frac{h^2\left(-3x^2+h^2\right)}{\left(x^2+h^2\right)^3}$$

$$\tau(x) = -b\sigma_{xy}^{D}(x,h) = -\frac{4Bb}{\pi}\frac{hx\left(h^2-x^2\right)}{\left(x^2+h^2\right)^3} \tag{A4}$$

Let u(x) denote the beam displacement due to bending. By convention, downward displacements and negative slopes are considered to be negative displacements and angles. The bend (tangent) angle $\phi = du/dx$ and the deflection $\delta = u$ (Fig. A3) are obtained by integration of the moment-curvature equation [6]

$$EI\frac{d^2u}{dx^2} = M(x) \tag{A5}$$

E is the elastic modulus and $I = bh^3/12$. The final results are independent of the beam width b and can be obtained in closed form as follows.

$$\phi(X) = -\frac{6B}{\pi Eh^2}\left[\tan^{-1}(X) + \frac{X}{1+X^2}\right] = \frac{6B}{\pi Eh^2}f(X)$$

$$\delta(X) = -\frac{6B}{\pi Eh}\left[X\tan^{-1}(X)\right] = \frac{6B}{\pi Eh}g(X) \tag{A6}$$

$X = x/h$ is the normalized distance along the plate (beam). The bend-angle and displacement functions, f(X) and g(X), defined in eq. [A6] are shown in Fig. A4; the deflection is symmetric with respect to the beam center, X = 0. The bend-angle function f(X) approaches the value $-\pi/2$ when X increases. The deflection function g(X) becomes essentially linear as f(X) approaches $-\pi/2$. This means that the bending effect appears as a localized "hinge" where the outer regions of the sample are simply carried downward by the bend-effect in the region containing the scribes.

The bending effect for multiple scribes across the width of a plate sample can be obtained from the superposition of the single-scribe functions in eq. [A6]. This assumes that the scribes are sufficiently widely spaced so as not to interact. If 2n +1 scribes are made on a plate such that n equally spaced scribes lie each side of the center scribe at X = 0, the bend angle and deflection functions will be as follows

$$\phi_{2n+1}(X) = \frac{6B}{\pi Eh^2}\sum_{j=-n}^{n} f(X+jA) = \frac{6B}{\pi Eh^2}(2n+1)f_{2n+1}(X)$$

$$\delta_{2n+1}(X) = \frac{6B}{\pi Eh}\sum_{j=-n}^{n} g(X+jA) = \frac{6B}{\pi Eh}(2n+1)g_{2n+1}(X) \tag{A7}$$

p is the spacing between adjacent scribes and A = p/h is the normalized scribe spacing. The average bend angle or displacement functions per scribe are defined as

$$f_{2n+1}(X) = \frac{1}{2n+1}\sum_{j=-n}^{n} f(X + jA)$$
$$g_{2n+1}(X) = \frac{1}{2n+1}\sum_{j=-n}^{n} g(X + jA) \tag{A8}$$

The general form of the functions $f_{2n+1}(X)$ and $g_{2n+1}(X)$ is shown in Fig. A4 for a single scribe, three scribes (n = 1) and five scribes (n = 2) with a scribe spacing A = 2/3 (p = 2h/3). When X increases, $f_{2n+1}(X)$ approaches the single-scribe function f(X) whereas $g_{2n+1}(X)$ is shifted to a smaller value than g(X). Because of this behavior, the bend angle has been used in this study for calibration of the dipole strength. The method adopted here to evaluate B was to make measurements of the average bend angle per scribe, which is defined to be the total bend angle divided by the number of scribe traces. As seen in Fig. A4, the bend angle function f_{2n+1} (X) asymptotes to a constant value for sufficiently long plate samples. This means that the outer regions of the beam are simple carried along as straight sections by the localized plastic hinge. Therefore, the opposing outer regions of bend profiles were fit to straight lines in order to calculate overall bend angles. The dipole strength parameter B is determined by equating the measured angle to the predicted angle given by eq. [A7]). If the sample L is too small to reach the straight-line asymptote for the bend-profile arms, the measured and predicted angles can be equated at a suitable mean measurement position [6].

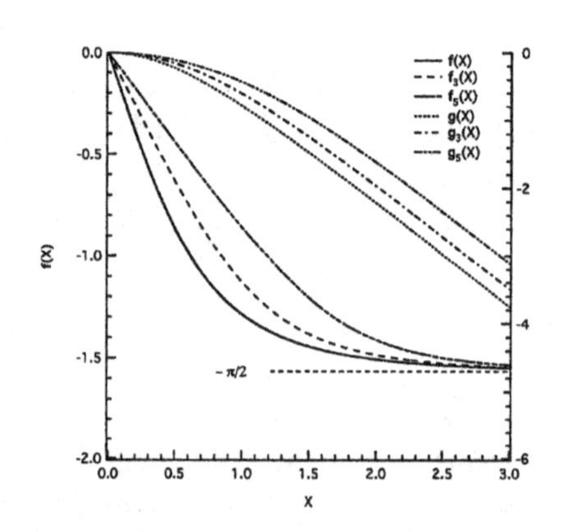

Figure A4. Plot of the normalized bend angle and deflection functions for a single scribe, f(X) and g(X), three scribes, $f_3(X)$ and $g_3(X)$ and five scribes, $f_5(X)$ and $g_5(X)$, using A = a/h = 2/3.

ACKNOWLEDGMENTS

The authors are indebted to Professor Jeff Eischen, North Carolina State University, for critical comments on the development the bend-effect model and also for the FEM verification calculations. The authors gratefully acknowledge the support provided by IBM, Data Storage Division, San Jose CA, and the NSF, Award No. DMR–0203552. The research was conducted at the Precision Engineering Center, North Carolina State University.

REFERENCES

1. R. F. Cook and G. M. Pharr, "Direct Observation and Analysis of Indentation Cracking in Glasses and Ceramics", J. Am. Ceram. Soc., 73 [4], 787 (1990).
2. B. Lawn, "Fracture of Brittle Solids", 2nd Edition, Cambridge Solid State Science Series (1993).
3. B. R. Lawn, "Indentation of Ceramics with Spheres – A Century after Hertz", J. Am. Ceram. Soc., 81 [8], 1977 (1998).
4. S. S. Chiang, D. B. Marshall and A. G. Evans, "The Response of Solids to Elastic-Plastic Indentation. I. Stresses and Residual Stresses", J. Appl. Phys., 53(1), 298 (1982).
5. E. H. Yoffe, "Elastic Stress Fields Caused by Indenting Brittle Materials", Phil. Mag. A, vol. 46, no. 4, 617 (1982).
6. B. W. Austin, MS Thesis, North Carolina State University, 2000.
7. Y. Ahn, T. N. Farris and S. Chandraskar, "Elastic Stress Fields Caused by Sliding Microindentation of Brittle Materials", Proc. of the Int. Conf. On Machining of Advanced Materials, S. Jahanimir, ed., NIST Spec. Publ. 847, 71 (1993).
8. R. O. Scattergood and U. F. Kocks, "Dislocation Pair Interaction in a Finite Body", NBS Special Publication 317, vol. II, 1179 (1970).
9. T. Randall, MS Thesis, North Carolina State University, in progress (2003).
10. W. S. Blackley and R. O. Scattergood. "Ductile-Regime Machining Model for Diamond Turning of Brittle Materials", Precision Engineering, [13], 95 (1991).
11. T. G. Bifano, T. A. Dow and R. O. Scattergood, "Ductile-Regime Grinding: A New Technology for Machining Brittle Materials", J. of Engr. for Industry [113], 184 (1991).
12. K. W. Sharp, M. H. Miller and R. O. Scattergood, "Analysis of the Grain Depth-of-Cut in Plunge Grinding", Precision Engineering, [24], 220 (2000).
13. C. Tam, C. C. Poon, L. Crawforth, and P. M. Lundquist, "Stress On The Dotted Line", Data Storage No. 6, 29 (1999).
14. B. M. Love, MS Thesis, North Carolina State University, 2001.
15. K. L. Johnson, "Contact Mechanics", Cambridge Univ. Press, p.17 (1985).

DUCTILE GRINDING OF BRITTLE MATERIALS

R. O. Scattergood
Materials Science and Engineering Dept.
North Carolina State University
Raleigh, NC 27695-7907

INTRODUCTION

Grinding of brittle materials is important for the manufacturing of components from these materials. Grinding of advanced ceramics can contribute 50 % or more to the total cost of a finished component. Control of the subsurface damage is one of the crucial factors for the machining operation since this can severely limit the performance of brittle materials. With the advent of high-stiffness, high-precision machine tools it is now possible to machine brittle materials without introducing subsurface fracture damage. This limit has been termed "ductile-regime" machining. The profile of the uncut chip in a machining operation is central to the understanding of the subsurface damage produced. Chip profile geometry will interplay with the material removal mechanisms such that the damage left in the part will be controlled by both the geometry and the material response. The latter can be characterized as a critical depth-of-cut for the onset of a ductile-to-brittle fracture transition. Also important is the size scale of the fracture damage produce at this transition. The factors underlying the ductile-regime machining concept are presented in this overview paper along with a brief discussion of test techniques that can be used to estimate the critical depth-of-cut. Single-point machining is used as a prototype for grinding because the geometrical aspects are most easily understood for this case. Plunge grinding tests provide insight into the extension of the critical depth-of-cut concepts to grinding.

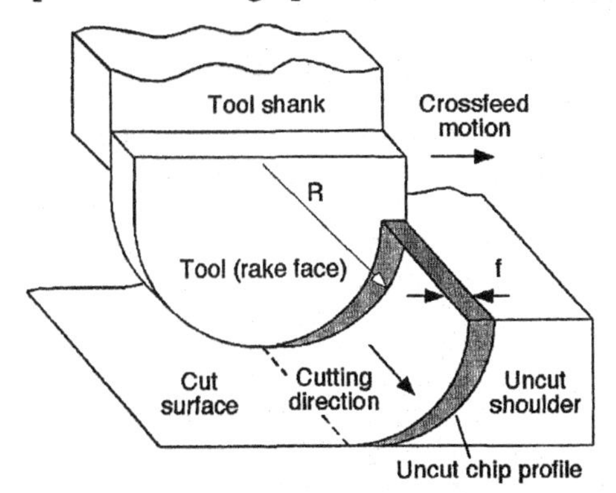

Figure 1. Chip profile geometry for single-point machining using a round-nose cutting tool.

SINGLE-POINT MACHINING

The essential connections between chip profile geometry, material removal mechanisms and the fracture damage introduced into a workpiece can be illustrated using single-point machining. For brittle materials, this type of machining is used to fabricate high-precision components for applications such as reflective mirrors and infrared detector optics. Since single-crystal diamond tools are usually needed to obtain the surface finish and form accuracy required, the process is often called single-point diamond turning. High-precision computer-controlled machining tools with positioning accuracy on the order of a few nanometers are required. It must be noted, however, that single-point machining, although quite successful for many classes of brittle materials, has limited applicability. Grinding is a more versatile process and will be the choice for high-volume manufacturing of advanced ceramics.

Fig. 1 shows the geometry for a facing cut made with a round-nose single-point cutting tool with nose radius R. Due to the cutting motion and the crossfeed motion of the tool, the uncut chip profile appears as the shaded section shown in Fig. 1. The feedrate f is the crossfeed distance traveled per workpiece revolution. It is important to note that this three-dimensional geometry provides a different perspective than the two-dimensional orthogonal cutting geometry that is commonly used to describe machining processes [1].

The effect of material response and its interplay with chip geometry can be illustrated using Fig. 2, which is a projection of Fig. 1 along the cutting direction. The chip profile is the sector bounded by two circular arcs defined by the tool nose radius R. Separation of the arcs is determined by the feedrate f. The effective depth-of-cut d varies from zero at the cut surface to a maximum value at the top of the uncut shoulder. For brittle materials, the extensive literature on indentation fracture [2] shows that there will be a transition from ductile plastic flow to fracture at a critical depth-of-cut d_c (plastic flow energy scales with volume while fracture energy scales with area, therefore, as the size scale of the deformation process increases, minimum energy input requires a transition from plastic flow to fracture at a critical depth). This can also be expressed as a critical load P_c, however, d_c is preferred because machining operations control the cutting depth rather than the load. Fracture damage with a characteristic size scale y_c will be initiated at d_c as shown in Fig. 2. This ductile-to-brittle transition is endemic to any machining or grinding process used for brittle materials such as glass, semiconductors or ceramics.

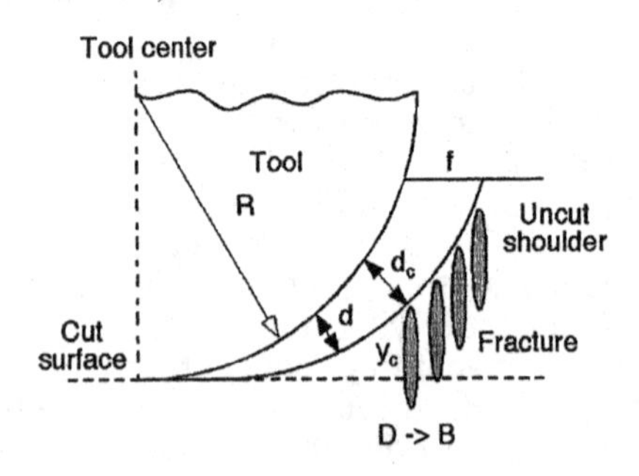

Figure 2. The ductile-to-brittle transition (D->B) that occurs at the critical depth-of-cut $d = d_c$.

The interplay between chip profile geometry and material response follows directly from Fig. 2. If the depth of fracture damage y_c extends below the plane of the cut surface, then the part will contain subsurface damage and its extent will depend upon the location of d_c along the chip profile boundary. The relevant machining parameter is the feedrate f, as opposed to the total depth of cut between the cut (finished) and uncut surfaces. Smaller values of f will move d_c higher along the chip profile. This effect will pull the lowest portion of fracture damage zone upward from the cut surface. There will be a maximum value f_{max} such that for $f < f_{max}$ the cut surface will remain damage free, even though a significant amount of material removal can still occur by fracture in the uncut shoulder. It is important to recognize the effect of chip profile geometry here. Changes in the chip profile due to changes in parameters such as feedrate f will control the subsurface fracture damage for given material response parameters d_c and y_c. These latter parameters will vary with material properties and tool loading conditions. High precision machines using air bearings are required for achieving ductile-regime machining in practice since the critical feedrate f_{max} will typically be on the order of $< 10\ \mu m$ for brittle materials. Machines with conventional bearings and lower stiffness will have error motions that are too large to maintain proper control of the federate at the accuracy required.

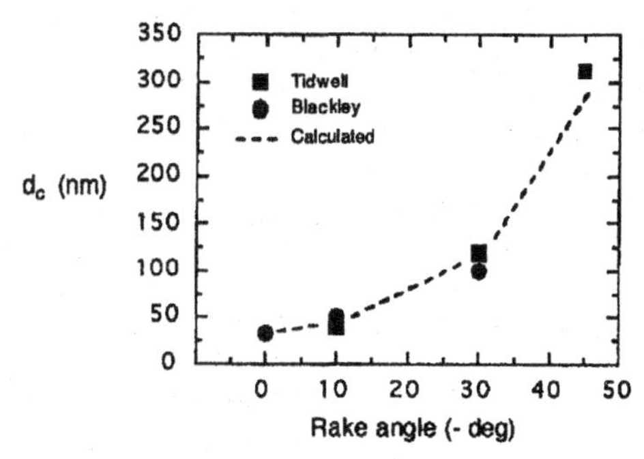

Figure 3. d_c vs. rake angle for machining of Ge. Data from [3] (circles) and [4] (squares). The dashed line is based a stress model described in [4].

Interrupted cutting tests using single-point diamond turning have been made to verify the predicted effect of the chip profile geometry shown in Fig. 2. For an interrupted cutting test, the cutting tool is rapidly withdrawn from the workpiece using a servo control and the position of the onset of the fracture damage zone on the uncut shoulder relative to the tool centerline can be measured. By conducting experiments over a range of feedrates, the parameters d_c and y_c can be obtained from the analysis of the experimental data [3,4]. It is necessary to account for damage accumulation in the shoulder region due to successive tool passes when doing this analysis. Fig. 3 shows measured values of d_c for (100) germanium crystals, machined along a [011] direction, as a function of the tool rake angle. d_c increases with increasing negative rake angle because of the change in stress loading conditions. Larger negative rake angles suppress fracture due to the larger compressive stress that develops with the increased tool thrust forces [5]. The results are also affected by crystallography (cutting plane and direction). Crystallographic fracture effects have been analysed for single-point machining of Ge and Si wafers by assuming that fracture damage inititates on the {111} cleavage planes [6].

The parameters d_c and y_c will determine the maximum feedrate for machining in the ductile regime, i.e., the value f_{max} at which the fracture damage in Fig. 2 first reaches the plane of the cut surface as f increases. For values of $f \le f_{max}$, there will be no subsurface fracture damage. Analysis of the chip profile geometry in Fig. 2 gives the relation [3,4],

$$f_{max} = d_c \sqrt{\frac{R}{d_c + 2y_c}} \tag{1}$$

R is the tool nose radius. The results from interrupted cutting tests have confirmed this relation. For given conditions, the depth of fracture damage is often proportional to the cutting depth, and y_c will be proportional to d_c. The critical depth d_c can then be viewed as a key material response parameter in eq. (1).

MICROSTRUCTURE EFFECTS

In higher toughness ceramics where the microstructure is designed to produce constraints that enhance toughness, for example, by means of crack bridging, fiber reinforcement or phase transformation, the toughening mechanisms may not be relevant for ductile-to-brittle transitions during machining. Toughening mechanisms require a certain measure of crack extension in order fully develop crack-tip shielding constraints. A large amount of research effort has been devoted to identifying these mechanisms and it is clear that toughness can be crack-size dependent. This is termed T-curve (or R-curve) behavior [7,8]. Fig. 4 shows toughness vs. crack size data for a glass ceramic toughened by heat treatment to produce a lenticular crystal phase. The data were obtained using measurements of crack extension under biaxial loading from Vickers indentation precracks [9]. Careful analysis of the test data is needed to obtain absolute values of the toughness using indentation precrack techniques, as has been discussed by many authors [10-12].

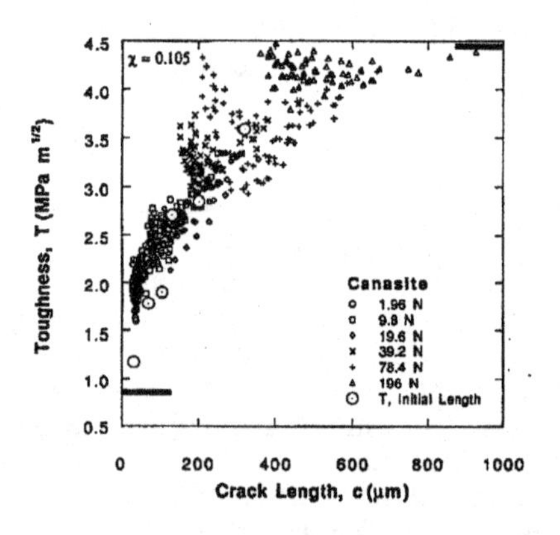

Figure 4. Fracture toughness vs. crack length for the glass ceramic Canasite [9].

A relationship between d_c and material properties can be developed using indentation fracture mechanics [2]

$$d_c = \beta \left(\frac{E}{H}\right)\left(\frac{K_c}{H}\right)^2 \tag{2}$$

K_c is the fracture toughness, H is the hardness, E is the elastic modulus and β is a constant depending upon indenter geometry and test conditions (indenter shape, sharpness, etc.). This relation was verified using acoustic emission techniques to monitor the onset of crack initiation during static indentation tests [13]. As will be discussed further in the section on plunge grinding, Bifano, et. al. [14] used a similar relation to correlate d_c values obtained for grinding tests. The onset of fracture for a ductile- regime machining operation typically occurs at crack size scales less than 10μm. The operative K_c value for "tough ceramics" in this size range will be significantly less than the long-crack peak value (Fig. 4). This means that d_c values in eq. (2) will be controlled by short-crack K_c values that are more indicative of the un-toughened base material. This point should not be overlooked when developing ductile-regime grinding operations for engineering ceramics.

SCRATCH TESTS

Static indentation tests for measuring d_c values do not include the cutting (tangential) force component present in machining. Parameters such as tool geometry or cutting speed cannot be taken into account using indentation tests. Development of scratch-type cutting tests has therefore been an important part of research for machining of brittle materials. In this type of test a cutting tool, which is often a Vickers or Rockwell diamond indenter, is driven along the workpiece under fixed loading conditions. Fracture damage occurs within the cut groove above a critical load. The general features of the fracture damage found for scratch tests resemble those found for machining.

Using a novel cross-sectioning technique [15], Jahanimir and coworkers [16,17] used scratch-type test techniques to study subsurface damage produced in commercial ceramics by a conical cutting tool (Rockwell indenter). Comparison with grinding tests on the same materials provided insight into the material removal mechanisms and confirmed the strong influence of T-curve effects on grinding performance in higher toughness ceramics. Tanikella et al [18,19] used a PZT actuator to produce a varying depth scratch by ramping a Vickers indenter cutting tool into a moving workpiece at a fixed loading rate, holding at a specified peak load, and then ramping out. The onset of fracture at a critical load was determined by acoustic emission monitoring. A more direct measure of d_c can be obtained if the

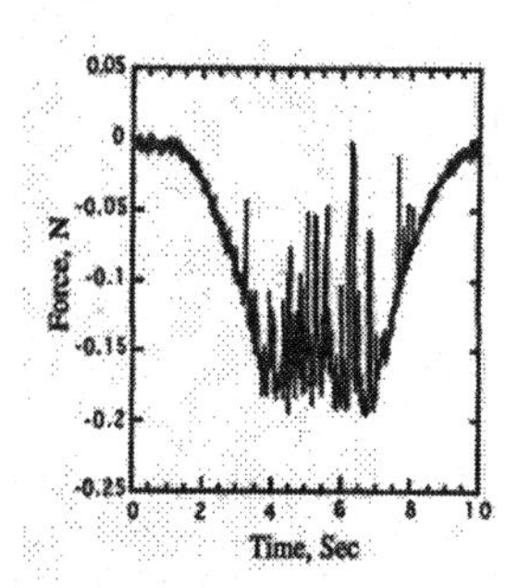

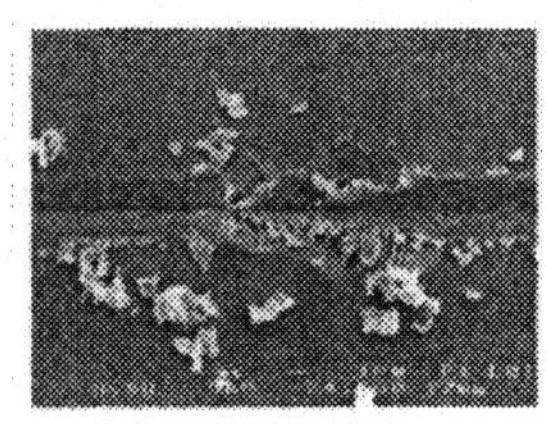

Figure 5. Top: Force profile for a controlled-depth scratch test. Middle: Ductile cutting region. Bottom: A "pop in" fracture event that corresponds to an unloading pulse in the force profile. [20,21].

cutting depth is controlled rather than the load [20,21]. Fig. 5 shows results obtained from controlled-depth scratch tests on silicon using a high stiffness force dynamometer to measure loads. The tests were done on a precision diamond turning machine by locking the spindle and controlling the cutting tool and workpiece motions with the slides (these motions are controlled to nanometer resolution using laser interferometers). Flatness of the sample was measured by profiling the sample with a capacitance gage and this data was used to subtract the sample slope from the desired tool path. The sharp unloading pulses on the force profile in Fig. 5 correspond to "pop-in" of individual fracture events. Between these events, ductile-regime cutting occurs. The first fracture event during the ramp-in part of the profile (left side) determines the d_c value for given cutting conditions. Although scratch tests are a very useful extension of static indentation tests, scratch tests are normally done at low speeds (mm/s or less), and therefore do not achieve the range of cutting speeds used for grinding (m/s).

PLUNGE GRINDING TESTS

A central issue for ductile-regime grinding is the relation between the chip profile and the effective depth of cut d. This relationship was easily visualized for single-point machining because the chip profile in Fig. 2 directly reflects the geometry of the single-point cutting tool profile. For grinding, the cutting tool (wheel) contains within a binder a distribution of grinding grains that produce multi-point cutting action. Fig. 6 shows the chip profile and ductile-to-brittle transition geometry for single-point machining in Fig. 2 modified such that the grinding wheel binder contains an embedded distribution of cutting grains. The cutting depth d used in single-point machining must be replaced by an average depth-of-cut per grain d_g. The latter will depend upon the number of active cutting grains that participate in material removal. This is determined by the grinding parameters and the distribution of grains on the wheel.

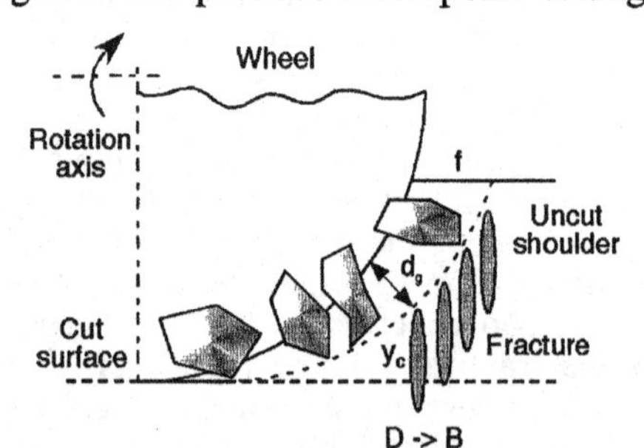

Figure 6. The ductile-to-brittle transition (D->B) for grinding that occurs at the critical grain depth-of-cut $d_g = d_c$.

The uncut chip profile for grinding is more complicated than that shown in the projection view in Fig. 6 because there is an upfeed motion due to the wheel rotation in addition to the crossfeed f [22,23]. However, the essential features for determining d_g for grinding can be analyzed using a simplified grinding process called plunge grinding shown in Fig. 7. In this case a rotating grinding wheel is plunged vertically into a stationary workpiece at a specified infeed f per revolution. If the wheel ran perfectly true about its rotation axis, the cutting action would be continuous. However, a grinding wheel can never be made to run perfectly true and there will be dynamical vibrations due to the rotating components. Runout effects of this kind produce an intermittent cutting action, similar to fly-cutting, and this can be taken into account for the determination of grinding chip profiles [22,23].

Analytical and computer simulation models were developed by Sharp et. al. [24] to determine the grain depth-of-cut d_g for plunge grinding. The chip profile for plunge grinding can be taken as the cross section fW shown in Fig. 7. f is the infeed per revolution of the wheel and W is the wheel width. The mean cutting depth per grain d_g, or equivalently the cutting area per grain A_g shown in Fig. 7, is determined from the steady-state requirement that $N_c A_g = fW$ where A_g is the average projected cutting cross section area per grain (relative to the cut surface at line aa) and N_c is the number of active cutting grains on one revolution of the wheel. This condition determines the d_g vs. f relation for plunge grinding. The results of the analytical modeling and computer simulations are shown in Fig. 8 (solid lines are analytical predictions and points are computer simulations). There are two regions for a d_g vs. f relation. At very low values of f, the relation is simply $d_g = f$. This region produces "correlated surfaces" such that each grinding groove on the final cut surface is produced by the cutting action of only one grain. As f increases in this region, the groove topology remains essentially unchanged while the grooves become uniformly deeper. At a certain transition point, the slope of the d_g vs. f curve in Fig. 8 (E = 0) decreases markedly. In this region the cutting action of the active grinding grains overlaps during one rotation of the wheel. Subsequent cutting grains will eliminate a groove formed by a leading grain. This "statistical cutting" action is most likely to be the operative one for production operations. Assuming no runout effects (E = 0) and an infinitely stiff binder containing the cutting grains, the d_g vs f relation for the statistical-cutting regime is [24]

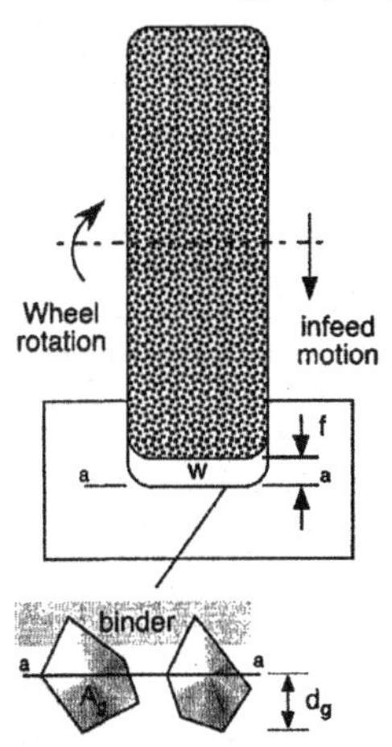

Figure 7. The geometry for plunge grinding is shown. f is the infeed per wheel revolution. Grinding for this mode produces a series of linear grooves on the ground surface (aa).

$$d_g = 0.587D\left(\frac{f}{F_v L}\right)^{0.4} \tag{3}$$

D is the mean diameter of the grinding grains, F_v is the volume fraction of grains in the binder and L is the grinding wheel circumference. The interplay of the grinding parameters is manifest in this relation. Ductile-regime grinding will be facilitated by reducing the d_g such that $d_g < d_c$. For a fixed wheel size (L constant), reducing the grain size or increasing the volume fraction of grains will favor ductile-regime grinding. Experience indicates that this is the case. If wheel runout is present, then the wheel need not make full engagement during one rotation, and fewer grinding grains are active. This increases the d_g value as is shown in Fig. 8 for the cases where the runout error E > 0. The effect of grain binder can also be included in the model, but these results are not shown in Fig. 8. With all else constant, a softer binder increases d_g since grains are "pushed" back into the wheel by the grinding thrust forces and so are less effective at material removal.

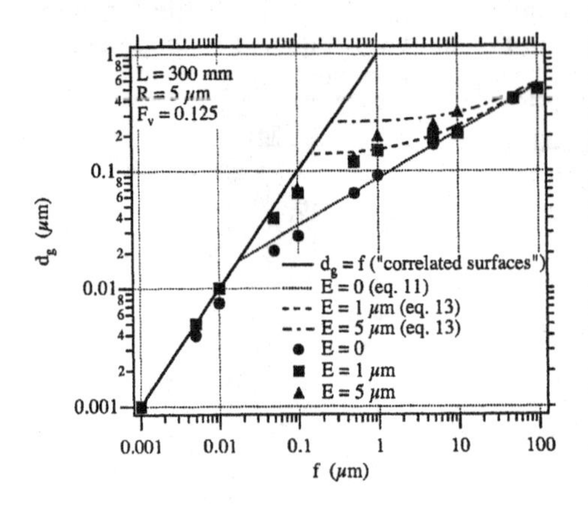

Figure 8. d_g vs. f curves obtained using a model for plunge grinding. E is the amplitude of the runout error and R is the grain radius. Equations are given in [24].

Plunge grinding can be used to measure d_c values for grinding. Using the results of the analysis presented in Fig. 8, the d_g value can be estimated from the infeed f for specified grinding conditions. This allows the critical value $d_g = d_c$ to be determined for the value of f corresponding to fracture initiation in the plunge-ground surface. The latter can be measured using optical microscopy observations as shown in Fig. 9, or with surface roughness measurement made along the groove directions [25]. Since the plunge-ground surfaces contain grooves formed by cutting grains, they will have a large roughness when measured across the groove direction.

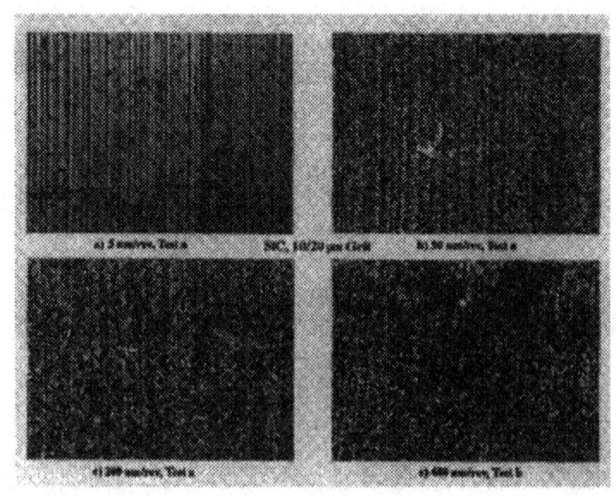

Figure 9. Optical micrographs for plunge-ground surfaces for SiC at increasing values of f. Plots of f vs. fracture pit density obtained from the micrographs are used to determine the ductile-to-brittle transition.

Bifano et al [14,26] used plunge grinding to measure d_c values for a series of brittle materials, including several commercial ceramics. The grinding conditions were fixed and a series of grinding tests were done for each material in conjunction with optical microscopy measurements to determine the critical feedrate at the onset of fracture. The results are shown in Fig. 10. The measured d_c values are plotted against calculated d_c values obtained from eq. (2) using $\beta = 1$. The length scale for calculated d_c values is therefore arbitrary. Material property values were obtained from published data or vendor supplied data. A correlation line of slope = 1 is drawn through the data points (the relative position of the correlation line would determine the β factor for the grinding conditions used). A good correlation results except for alumina (point 9) and toughened zirconia (point 10). The fracture toughness values used for these two cases correspond to the peak (long crack) toughness value quoted for each material by the supplier. This clearly overestimates the d_c value because of the T-curve effects discussed in conjunction with Fig. 4. The horizontal line in Fig. 10 shows the range of d_c values obtained from interrupted single-point diamond turning tests on Ge using various machining conditions and tool rake angles (Fig. 3). The grinding result (point 6) falls within this range.

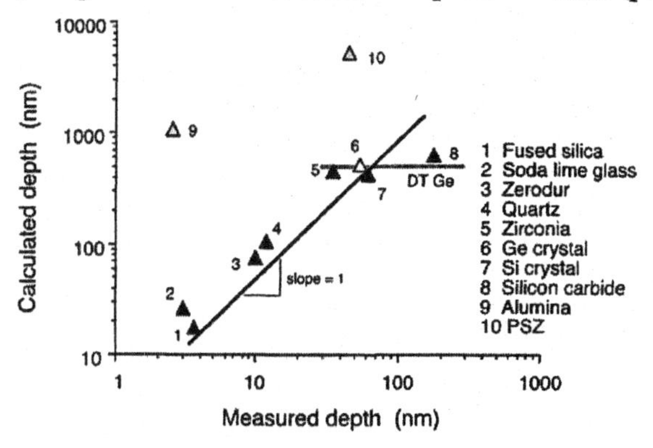

Figure 10. Plot of the calculated vs. measured values of d_c for plunge-grinding tests. The calculated values contain the scaling factor β in eq. (2).

SUMMARY

Material response characterized by a critical depth-of-cut leading to a ductile-to-brittle transition can, in conjunction with the chip profile geometry, be used to control the generation of subsurface fracture damage during the machining and grinding of brittle materials. This approach provides a rational means to develop process models for machining and grinding operations.

REFERENCES

1. G. Boothroyd, “Fundamentals of Metal Machining and Machine Tools”, McGraw-Hill (1975).
2. B. Lawn, “Fracture of Brittle Solids”, 2nd edition, Cambridge Solid State Science Series, Cambridge Press (1993).
3. W. S. Blackley and R. O. Scattergood, “Ductile-Regime Machining Model for Diamond Turning of Brittle Materials”, Precis. Engr., **13** [2] 95 (1991).
4. R. M. Tidwell, “Ductle Regime Machinine of Germanium: Development of New Experimental and Analytical Analysis Methods”, MS Thesis, North Carolina State University (1991).
5. K. Ueda, T. Sugita and H. Tsuwa, “Application of Fracture Mechanics in Micro-Cutting of Engineering Ceramics”, Annals of the CIRP, **32** [1] 83-86 (1983).
6. W. S. Blackley and R. O. Scattergood, "Crystal Orientation Dependence of Machining Damage - A Stress Model", J. Am. Ceram. Soc., **73** [10] 3113 (1990).
7. R. F. Cook, B. R. Lawn and C. J. Fairbanks, “Microstructure-Strength Properties in Ceramics”, J. Am. Ceram. Soc., **68** [11] 604-15 (1985).
8. S. J. Bennison and B. R. Lawn, “Role of Interfacial Grain-Bridging Sliding Friction in the Crack-Resistance and Strength Properties of Non-Transforming Ceramics”, Acta Metall., **37** [10], 2659-72 (1989).
9. S. M. Smith and R. O. Scattergood, "Short-Crack Toughness Curves in Ceramics", J. Am. Ceram. Soc. 79 [1], 129 (1996).
10. L. M. Braun, S. J. Bennison and B. R. Lawn, “Objective Evaluation of Short-Crack Toughness Curves Using Indentation Flaws”, J. Am. Ceram. Soc., **75** [11] 3049-57 (1992).
11. R. F. Krause, “Flat and Rising R-Curves for Elliptical Surface Cracks from Indentation and Superposed Flexure”, J. Am. Ceram. Soc., **77** [1], 172-78 (1994).
12. Bleise and R. F. Steinbrech, “Flat R-Curve from Stable Propagation of Indentation Cracks in Coarse-Grained Alumina”, J. Am. Ceram. Soc., **77** [2] 315-22 (1994).
13. J. Lankford and D. L. Davidson, “The Crack Initiation Threshold in Ceramic Materials Subject to Elastic/Plastic Indentation”, J. Mater. Sci., **12** 1662-68 (1979).
14. T. G. Bifano, T. A. Dow and R. O. Scattergood, “Ductile-Regime Grinding: A New Technology for Machining Brittle Materials”, Jour. of Engr. for Industry, **113** 184-89 (1991).
15. F. Guiberteau, N. P. Padture and B. R. Lawn, “Effect of Grain Size on Hertzian Contact Damage in Alumina”, J. Am. Ceram. Soc., **77** [7] 1825-31 (1994)
16. H. H. K. Xu, S. Jahanimir and Y. Wang, “Effect of Grain Size on Scratch Interactions and Material Removal in Alumina”, J. Am. Ceram. Soc., **78** [4] 881-91 (1995).

17. H. H. K. Xu, N. P. Padture and S. Jahanimir, "Effect of Microstructure on Material-Removal Mechanisms and Damage Tolerance in Abrasive Machining of Silicon Carbide", J. Am. Ceram. Soc., **78** [9] 2443-48 (1995).
18. B. V. Tanikella, K. A. Gruss, R. F. Davis and R. O. Scattergood, "Indentation and Microcutting Fracture Damage in a SiC Coating on Incoloy Substrates", Surf. and Coatings Tech., 88, 119 (1996).
19. B. V. Tanikella and R. O. Scattergood, "Fracture Damage in Borosilicate Glass During Microcutting Tests", Scripta Met. et Mater. 34 [2], 207 (1996).
20. B. V. Tanikella, "Controlled Microcutting Tests on Glassy Materials and Single Crystal Silicon", PhD Thesis, North Carolina State University (1996).
21. B. V. Tanikella and R. O. Scattergood, "Controlled Microcutting Tests on Single Crystal Silicon", Proc. ASPE Spring Topical Meeting, Annapolis MD (1996).
22. S. C. Fawcett and T. A. Dow, "Influence of Wheel Speed on Surface Finish and Chip Geometry in Precision Contour Grinding", Prec. Engr., **14** [3] 160-67 (1992).
23. G. E. Storz and T. A. Dow, "Cup Wheel Grinding Geometry", Proc. ASPE Annual Mtg., Cincinnati, Ohio, 105 (1994).
24. K. W. Sharp, M. H. Miller and R. O. Scattergood, "Analysis of the Grain Depth-of-Cut in Plunge Grinding", Prec. Engr., 24, 220 (2000).
25. K. W. Sharp, "Development of Grinding Models for Brittle Materials", PhD Thesis, North Carolina State University (1998).
26. T. G. Bifano, "Ductile Regime Grinding of Brittle Materials", PhD Thesis, North Carolina State University (1988).

NANO-COMPRESSION OF CARBON MICRO-BALLOONS WITH A FLAT-ENDED CYLINDRICAL INDENTER

G. Gouadec*
LADIR - U.P.M.C.
2 rue H. Dunant
94 320 Thiais, France

K. Carlisle, K.K Chawla and M.C. Koopman
School of Engineering
University of Alabama at Birmingham
Al, 35205

G. M. Gladysz, M. Lewis
Eng. Sci. & Applications Division
Los Alamos National Laboratory
NM, 87545

ABSTRACT

We report on what is, to our knowledge, the first experimental compression campaign carried out on individual micro-balloons. Single Carbon Micro-Balloons (CMBs) were compressed between 5 and 50 mN on an XP nano-indentation device (MTS) customized with a special cylindrical tip. Of the 140 tested CMBs, approximately 22% were classified as single-walled spherical microballoons, on the basis of geometrical and mechanical criteria. The remaining exhibited either non spherical morphologies and/or defects in the structure. Details of this classification are presented. Compression testing revealed that the CMB displacement was proportional to the square root of the microballoon diameter. Conventional Berkovich nano-indentation performed on debris from crushed CMBs yielded a Young's modulus of approximately 35 GPa for. The measured MB wall thickness was between 0.3 and 2.2 μm, showing no correlation with the diameter of the pristine CMB.

* Worked on the project as a post-doctoral student at the University of Alabama at Birmingham.

INTRODUCTION

Syntactic foams consist of thin hollow spheres called Micro-Balloons (MBs) held together with a binder phase. Binder phases of ceramic, metal and polymer are reported in literature. In general, the materials are attractive for their very high specific strength (stems from the possibility for thrust forces to develop in curved surfaces [1]). Applications of syntactic foams include submersed equipment, reinforcing materials for sandwich composites and, because of the high void content, low dielectric constant substrates [2].

Like any composite structure, mechanical properties of the MBs -the reinforcement- are integral to those of the syntactic foams. Knowing them would be useful in modeling the system This paper presents nano-compression and nano-indentation results obtained on individual carbon MBs (CMBs).

EXPERIMENTAL

Materials

Phenolic microballoons, the precursor material for the CMBs, were supplied by Asia Pacific Microspher (Malaysia). Honeywell ALS (USA) converts the precursor to carbon through a proprietary carbonization process. Figure 1 shows typical SEM micrographs.

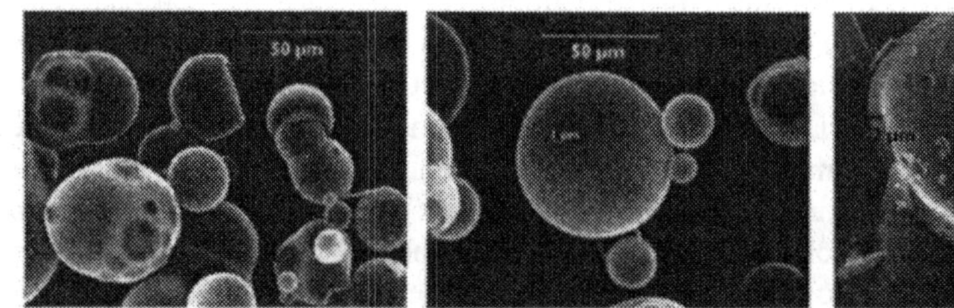

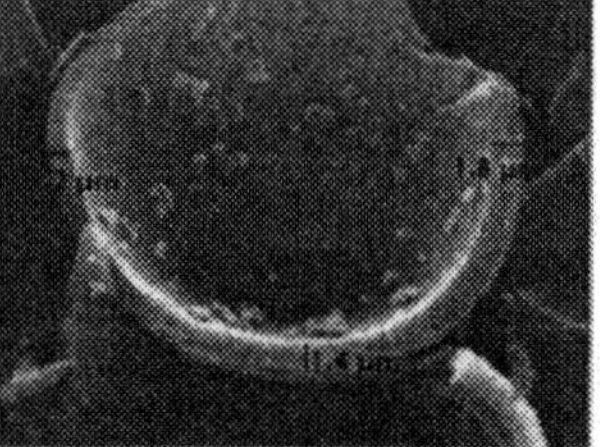

Figure 1: SEM micrographs of Honeywell-ALS Carbon Micro-Balloons (CMBs).

Nano-Compression Procedure

The compression machine was a displacement-controlled XP2 Nanoindenter from MTS-Nanoinstruments (Oak Ridge, TN). It has a maximum load of 500 mN (with 500 nN resolution) and a displacement resolution better than 0.02 nm. The calibration was checked on silica standard and produced mean Young's modulus and hardness values of 70 and 9.25 GPa, respectively.

In order to prevent wall puncturing, as was observed in a preliminary compression attempts with a Berkovich-shaped indenter, a flat-ended 89-μm diameter sapphire cylinder tip (special order from MTS) was used.

CMBs were evenly dispersed on polished stubs of Al6061 T651 alloy that were already set in the XP2 sample holder. A 40X magnifying lens coupled with the XP2 crosshead allowed diameter measurements of the CMB and aimed the indenter tip prior to compression. The average of two diameter measurements in orthogonal directions will hereafter be referred to as Φ_o (the subscript standing for "optical").

From the micrographs in Figure 1, large diameter and/or wall thickness fluctuations were observed from one CMB to another. CMBs (Φ_o from 3 to 60 μm) were compressed at an indentor tip velocity of 50nm/s. In spite of their significant abundance, we deliberately excluded the agglomerated CMBs visible in Figure 1.

Nano-Indentation Procedure

In an attempt to estimate the thickness and modulus of CMB walls, fragments left after nano-compression experiments were subsequently indented with a conventional three-sided Berkovich diamond (α=65^0; β=120^0). Being equipped with a Continuous Stiffness Measurement (CSM) capability, the XP2 Nanoindenter allows monitoring Young's modulus during the indentation, including the loading segment [3].

RESULTS AND DISCUSSION

Nano-Compression

Figure 2 shows typical compression curves. The first family was named "St" which corresponds to CMBs exhibiting steady loading prior to fracture. The "I" category regroups all balloons with an "imperfect" loading curve. As for the third type, "Sq", it is assigned to CMBs showing sequential fracturing. The displacement at rupture, δ_{max} (μm), the compressed diameter Φ_c (μm) and the load at rupture, L_{max} (mN), that are shown in Figure 2 were used to define the compression strain $C^{\%}$ (in %), the loading slope, S_L (in N/mm), and the aspect ratio, (A.R.), of compressed CMBs:

$$C^{\%}=100\times\frac{\delta_{max}}{\Phi_c} \tag{1}$$

$$S_L=\frac{L_{max}}{\delta_{max}} \tag{2}$$

$$A.R.=\frac{\Phi_c}{\Phi_o} \tag{3}$$

Only CMBs with A.R. in the range 0.9-1.1 will be assumed to be spheres*. Table I gives the distribution of the 140 tested CMBs.

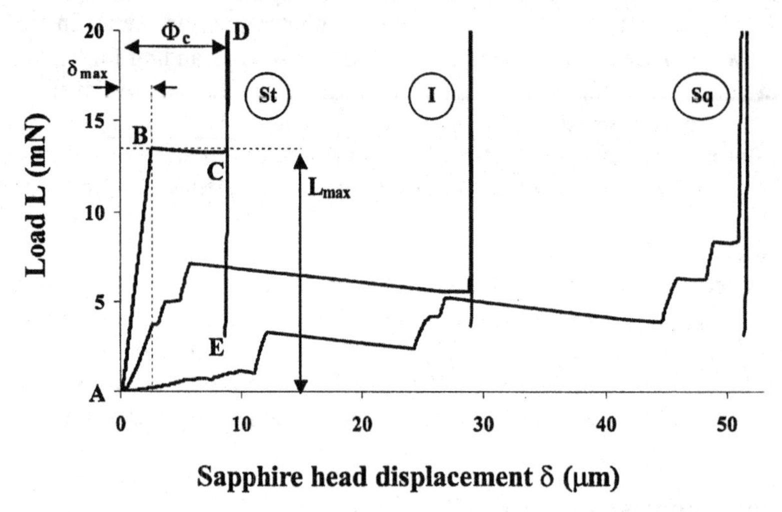

Figure 2: Compression curves of CMBs (St: Steady ; I: Imperfect ; Sq: Sequential) ; [AB]: steady compression ; B and C are consecutive points (the compression head moves to the substrate when the CMB fractures) ; [CD] and [DE]: loading-unloading of the aluminum alloy substrate.

Table I. Distribution of the different types of nano-compressed CMBs.

Geometry	Compression	Population	Percentage
Sphere	Steady	31	22
	Imperfect	6	4
	Sequential	7	5
Ellipsoid	Steady	54	39
	Imperfect	23	16
	Sequential	19	14

* Φ_o is an average of two values that were usually close to each other. Yet, $(\Phi_o)_{max}/(\Phi_o)_{min}$ exceeded 1.1 for 11 of the CMBs with an A.R. in the range [0.9-1.1], which will therefore not be counted as spheres.

Figure 3 is an optical micrograph of a three-phase CMB-reinforced syntactic foam that was manufactured and provided by Honeywell FM&T (the polishing technique was developed with the help of Struers Application Laboratory). This photograph reveals the existence of some compartments in the balloons and these unexpected flaws help to understand "Sq" curves. They simply correspond to the succession of "Steady-like" compressions of the different compartments. As for "T" curves, their "jumps" might indicate the sudden breakage of small compartments like those nested in the walls but could alternatively correspond to some "readjustment" of the CMBs position under the sapphire head. From now on, all sequential compressions will be disregarded.

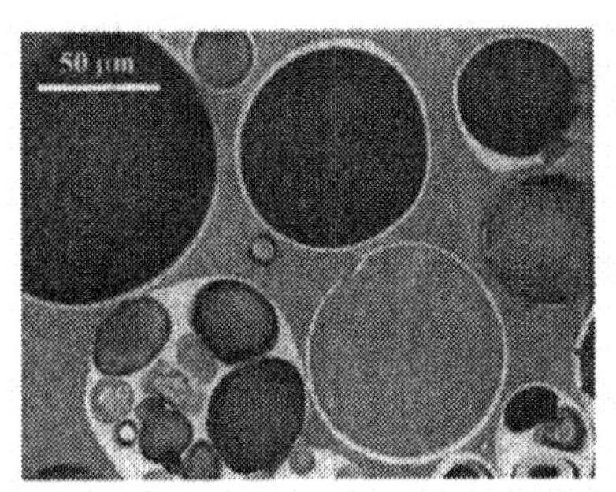

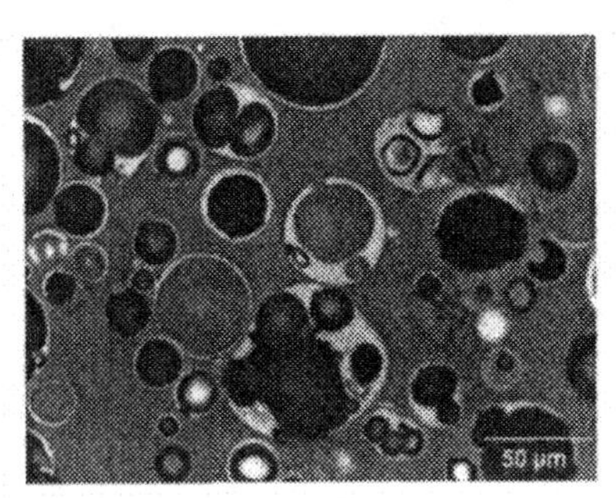

Figure 3: Optical micrograph of a resin-impregnated CMB foam.

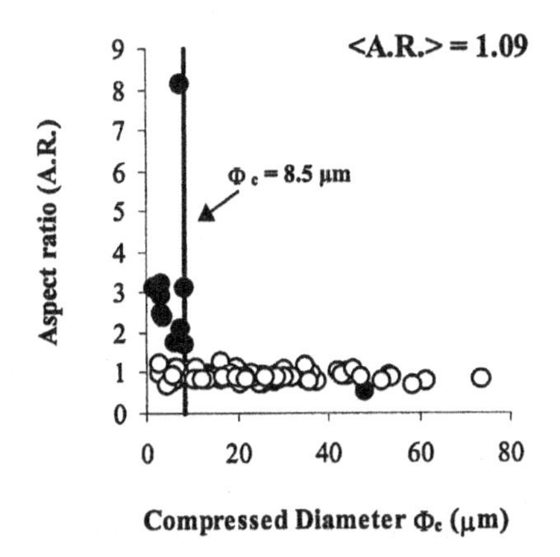

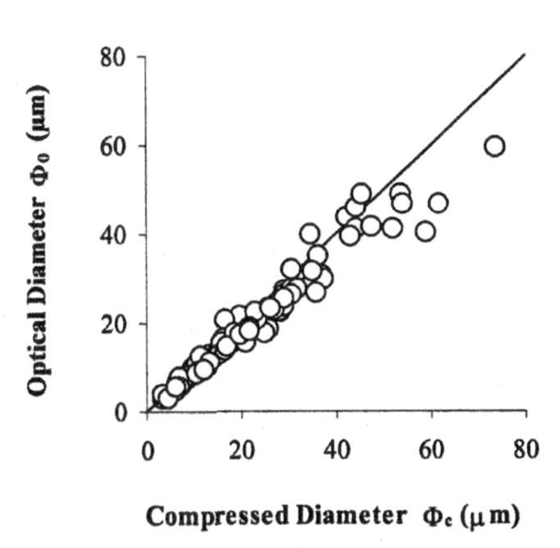

Figure 4: (left)- Aspect ratio of the 114 "St" and "T" CMBs as a function of the compressed diameter. Plain circles correspond to A.R. outside [0.7-1.3] ; (right)- Φ_0 vs. Φ_c plot for CMBs with an A.R. in [0.7-1.3].

Very few mechanical models are available in the literature for syntactic foams. Lee and Westmann [4] found an explicit formula for the bulk modulus as a function of the bulk and shear moduli of the matrix and MBs. Bardella and Genna [5] derived a more sophisticated model on the basis of a homogenization technique, taking the matrix porosity and the statistical distribution of the diameters and thickness of the MBs into account. These models are based on the common assumption that MBs are spherical. It is obvious from Figure 4(b) that CMBs with a compressed diameter less than 8.5μm (40% of them have A.R. ≥ 1.69) or larger than 50 μm have a tendency to be ellipsoidal. Non-spherical CMBs would be a hindrance to finding "behavioral laws" and therefore those with an A.R. outside of 0.7-1.3 will be disregarded. The remaining 103 CMBs and their compression data are given in Table II. The average A.R. is 0.95 for the smallest nineteen ($\Phi_c < 10$ μm) while it drops down to 0.85 when the biggest 8 are considered ($\Phi_c > 45$ μm). All in all, CMBs with an A.R. of just above 8.5 μm are the most likely to be "perfect" hollow spheres (steady compression, A.R. close to 1).

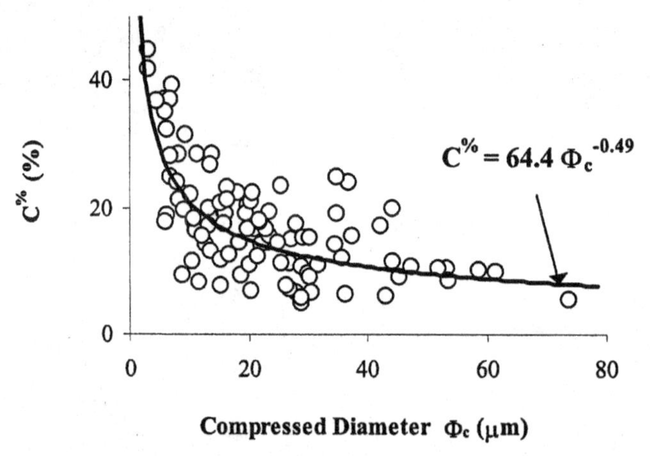

Figure 5: Compression strain $C^{\%}$ as a function of the Compressed Diameter. Only "St" and "I" CMBs with A.R. in the range [0.7-1.3] were considered.

Figure 5 shows experimental values of C% obey a $(\Phi_c)^{-y}$ law. There is a large dispersion, which is probably due to the random thickness of the walls, but y superscript is approximately ½. If the latter value is significant, then definition (1) implies:

$$\delta_{max} \propto \sqrt{\Phi_c} \tag{4}$$

If the compression behavior was governed by geometry only, then δ_{max} would be "statistically" proportional to Φ_c. The square root law in equation (4) suggests failure occurs earlier because of flaws, the probability of finding one such flaw increasing with Φ_c.

Table II. Statistics on "St" and "I" CMBs with AR in the range [0.7-1.3].

Parameter	Symbol	Unit	Min.	Max.	Mean	St. Dev.
"Optical" diameter	Φ_o	µm	3	59.5	20.2	12.1
"Compressed" diameter	Φ_c	µm	3	73.7	22.5	14.0
Fracture Load	L_{max}	mN	1.3	63.1	11.7	10.3
Loading slope	S_L	$N.mm^{-1}$	0.5	24.7	4.1	3.8
Compression strain	$C^{\%}$	%	5	44.8	17.5	8.6

Surprisingly enough, considering the geometrical specificity of the sample, there is still a proportionality factor between load and displacement (S_L) in a CMB nano-compression experiment (Figure 2, "St curve"). The data suggest that the CMBs are obeying Hooke's Law and that a "pseudo-modulus" E_{CMB} can be defined using suitable effective values for the stress (σ_{eff}) and deformation (ε_{eff}):

$$E_{CMB} = \frac{\sigma_{eff}}{\varepsilon_{eff}} \tag{5}$$

Assuming all of the load L is transmitted through the walls, noting R_{h-s} the sapphire head / CMB contact radius and t the wall thickness, the the simplest expression available for ε_{eff} ($\frac{2R_{h-s}}{\Phi_c}$), yields:

$$E_{CMB} \approx \frac{S_L}{2\pi} \times \frac{1}{t} \tag{6}$$

E_{CMB} calculated using the average values of S_L (Table II) and t (Fig. 7) is close to 0.6 GPa. This model* is very crude yet E_{CMB} calculated using equation (6) for CMBs

* will be developed in a forthcoming paper. The main approximation is to consider $\delta << 2\Phi_c$.

E_{CMB} calculated using the average values of S_L (Table II) and t (Fig. 7) is close to 0.6 GPa. This model* is very crude yet E_{CMB} calculated using equation (6) for CMBs of known wall thickness would be a possible tool to assess for the relation between CMB geometry and the mechanical properties of the syntactic foam.

Nano-Indentation

Figure 6 is an example of an indentation curve on a CMB wall and an illustration of how wall thickness was determined. The machine automatically computes Young's modulus. Only 13 of the 36 curves we recorded could be used to measure the thickness t with confidence. All other indentation curves were "sequential", starting with an irregular zone of small loads. This probably corresponds to some "rolling" of the debris under the Berkovich indenter, a "position adjustment" which made it difficult to detect initial contact of the indentation. Results are summarized in Figure 7.

A Φ^{-2} law would apply if all CMBs resulted from liquid droplets of the same volume V (at least for truly hollow, non compartmented, CMBs). Since the measured wall thickness showed no correlation with the diameter of the pristine CMBs, the modeling of the CMB and, consequently CMB-foams, is complicated.

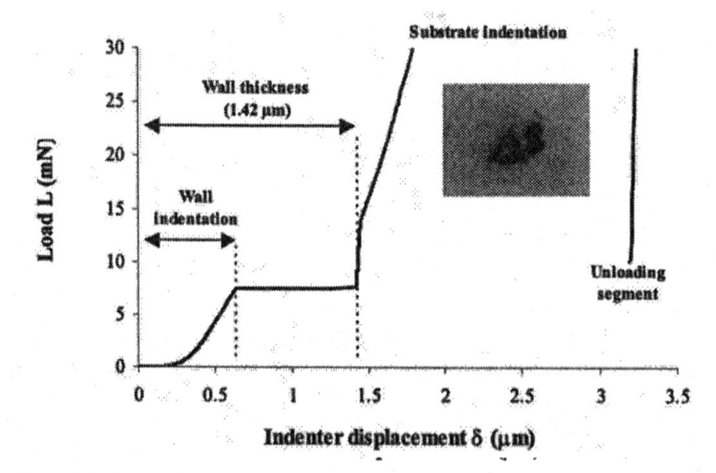

Figure 6: Nanoindentation of a compressed CMB wall. The insert shows the substrate alloy with the triangular print of the Berkovich indenter and a black debris of the indented fragment

* will be developed in a forthcoming paper. The main approximation is to consider $\delta \ll 2\Phi_c$.

Note our indentation values of t complement SEM results (Figure 1), even considering the tendency of the SEM to overestimate t when the wall section is not perpendicular to the line of sight. Nanoindentation is presumably more precise and allows for comparison between t and Φ but, unfortunately, few values can be obtained per CMB.

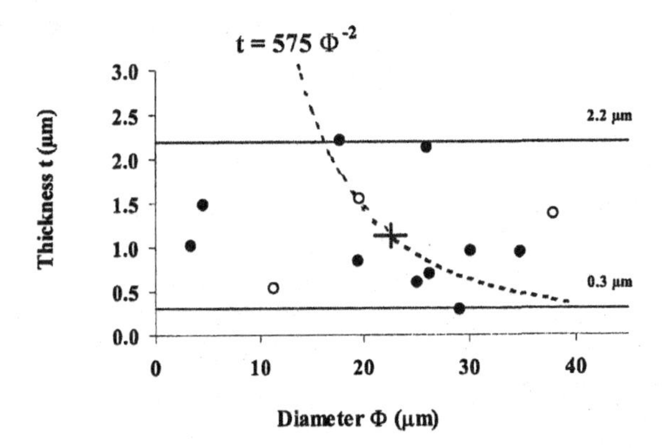

Figure 7: Measured thickness of the wall debris as a function of Φ_c (plain circles) and Φ_o (hollow circles). The dotted line is a Φ^{-2} law passing through the average point marked with a cross {Φ=22.5µm (Table II); t=<t>=1.12 µm}.

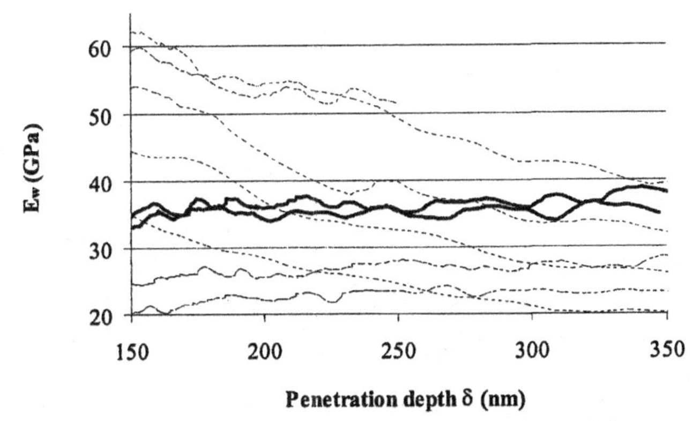

Figure 8: Modulus of indented wall fragments as a function of the penetration depth (CSM-equipped instrument).

Figure 8 gives measured values for Young's modulus of the CMB wall. In this case, not only is the indenter imperfect but also the sample surface is rough and curved. It may take a larger δ than the 150 nm from which we started plotting to get good tip/sample physical contact. Poor contact is probably the reason why most curves decrease. Furthermore, modulus computation implicitly assumes that the section of the indenter reaching the surface of the sample is smaller than the sample total area. This was not verified for a number of the fragments we indented, which meant a risk of underestimating E_W.

All in all, the most reliable curves in Figure 8 are the "constant modulus" curves of higher value: the wall modulus in CMBs is approximately 35 GPa.

CONCLUSION

Nano-compression testing was used to get a statistical view on the geometry (diameter, thickness) and micro-mechanics (fracture load and compression strain) of carbon micro-balloons (CMBs). Compartments could be observed in many CMBs by optical microscopy and account for the observation of "sequential" compression curves. In addition, the microballoon compression displacement at failure was proportional to the square root of the MB diameter, suggesting that the larger the MB diameter, the greater the probability of finding a flaw. The wall modulus was estimated to be 35 GPa with conventional Berkovich indentation. We also defined a pseudo modulus reflecting the mechanical stiffness of CMBs.

ACKNOWLEDGMENTS

The authors wish to thank A. Rossillon (UAB) and E. Herbert (MTS) for their precious help during this study. We also would like to acknowledge Los Alamos National Laboratory for funding this project under subcontract #44277-SOL-02 4X.

REFERENCES

[1] P.L. Gould, *Analysis of Shells and Plates,* 1999, Prentice Hall, Upper Saddle River, NJ 07458.

[2] J.K. Cochran, "Ceramic Hollow Spheres and their Applications," *Current Opinion in Solid State & Materials Science*, **3** 474-79 (1998).

[3]. B.N.Lucas, W.C. Oliver and J.E. Swindeman, "The Dynamics of Frequency-Specific, Depth-Sensing Indentation Testing," *Mat. Res. Soc. Symp. Proc*, **522** 3-14 (1998).

[4]. K.J. Lee and R.A Westmann, "Elastic Properties of Hollow-Sphere-Reinforced Composites", *J. Composite Materials*, **4** 242-53 (1970).

[5] L. Bardella and F. Genna, "On the Elastic Behavior of Syntactic Foams", *Inter. J. Sol. & Struct.*, **38** 7235-60 (2001).

OPEN FORUM: VICKERS INDENTATION TECHNIQUE–PROBLEMS AND SOLUTIONS

Ahmad G Solomah
SAC International Ceramics, Mississauga, Ontario L4W 4N5 CANADA

ABSTRACT

Problems associated with the application of Vickers indentation technique on brittle solids are being discussed and analyzed. The most serious problem encountered during our two decades of research on this very interesting testing technique is the surface residual stresses retained as a result of grinding and finishing during specimen preparation. Recommendations to overcome such a serious problem are being presented. Another serious problem is the use of the empirical formulae available in the literature, which were developed to calculate the toughness from indentation load, P, and indentation crack length, c, for the so-called "well-behaved" brittle materials. These empirical formulae have shown their inconsistencies in determining the toughness of several engineering ceramic materials including transformation-toughened zirconia-based ceramics. It is recommended to develop a new model that can accommodate such a phase transformation and any other chemo-thermodynamic changes within the material during indentation process. Other problems include microstructural features, e.g., secondary phase(s) in ceramic matrix composites; edge and neighboring indentation interactions and crack shape and crack length measurements, are also presented and discussed. Effects of environment on crack length measurement have been also discussed. Summary of discussions that took place during the sessions of the symposium are also included.

INTRODUCTION

Indentation techniques have been widely applied in the characterization of many engineering ceramic materials, e.g., from bioceramics to thin film in microelectronic devices and nano-composites. They possess attractive features like simplicity, economic factors and they can be performed on small specimens where other tests can not be used. With the rapid advancements in new materials development, the needs for sophisticated and reliable testing procedure increase in order to evaluate such materials for the intended applications. Since the introduction of Vickers indentation technique to determine the toughness of brittle solids by Evans and Charles in 1976 (1), using the crack length measurement and the indentation load, the technique became a routine method to measure the toughness of ceramic materials by almost every ceramic scientist. Evans and Charles used the mathematical analyses of the expanding cavity theory to develop a simple "empirical" formula to calculate the toughness within ~ 30% accuracy, when compared with the values measured using "standard" methods, i.e., DCB and SENB. Over the next two decades, many researchers tried to "improve" the above mentioned formula and to get "better" toughness values for many ceramics and brittle solids. They used the "raw data" of Evans and Charles or other published data that contain toughness values of several ceramic materials measured by other traditional techniques. Many non-linear data fitting were employed and about a dozen of new formulae were developed and applied by other researchers. Discrepancies among the "new" formulae were noticed and confusion occurred among researchers: which formula is the "appropriate" and more "realistic" one that can provide us with a "reliable" toughness value for such a test of a specific material? Empirical formulae that are mentioned above represent just one of several problems that are associated with indentation technique and its applicability to determine the toughness of brittle materials. The following sections describe the problems associated with the application of Vickers indentation technique in the characterization of brittle ceramic materials.

PROBLEMS

Sample Surface Preparation and Finishing

The surfaces of the brittle materials are often not representative of the bulk materials. The most serious problems encountered with surfaces are those related to residual stresses due to grinding and surface finishing, possible chemistry changes due to contamination and different microstructure as a result of prior processing, powder/chemical compositional heterogeneity, green body formation, etc. Unless indentation is intended to be used to study these surface characteristics/effects, they must be removed or a "standardized" treatment must be developed and followed which will guarantee create reproducible surface conditions. Research on transformation toughened zirconia ceramics, grinding and polishing have been the prime reason in producing erroneous results due to the residual surface stresses, i.e., compressive stresses. These compressive stresses impede the crack propagation and, subsequently shorter cracks were measured resulting in a superficial higher toughness (2). A "standardized' grinding and polishing procedures were developed in our laboratory to produce zirconia-toughened ceramic samples with a mirror-like surface finish with no residual stresses, thus eliminating the ambiguity in crack length measurements(3). Earlier studies have shown similar effects on hard metal alloys, e.g., WC- 6wt.%Co.

Indentation Load Effect on Crack Length

The original studies of Palmqvist on hard metals suggested that a linear relationship between indentation load, P, and crack length, l, where $l = c - a$, a is the half indentation diagonal, and c is the crack length measured from the center of indentation impression to the crack tip. Recent analyses predict that a linear relationship between P and $c^{3/2}$ should exist, assuming a half-penny crack shape. Some brittle solids behave in this way, particularly when c is much lager that a. The toughness of the brittle material can be determined from an equation that can incorporate the indentation load, P, and the crack length, c. A general relationship in the following form has been proposed by many researchers:

$$K_{IC} = \text{constant} \cdot E^m \cdot H^n \; P/c^{3/2} \quad \ldots\ldots\ldots\ldots\ldots (1)$$

where m and n are constants whose values are less than 1. E is the elastic (Young's) modulus and H is the hardness. Subsequently, several formulae were developed and used by almost every researcher to determine the toughness of materials under investigation. Discrepancies in the toughness values of a specific material were noticed when these "empirical" formulae were applied, as it has been demonstrated in a paper published in this Transaction (4). It was also observed that although a linear relationship between P and $c^{3/2}$ exists, the plot does not pass through the origin of the axes (see Figures 1 and 2 in reference 4). This observation requires that an adequate determination of the relationship between P and c.

The Geometry of the Indenter

The geometry of the pyramid indenter can have a significant effect on the indentation crack length. For example, changing the indenter angle from 136°–140° caused a decrease in *l* from 70–60 μm at a load of 50 kg on a WC-11 wt% Co hard metal alloy. Using an indenter with an angle of about 130° produced the maximum indentation crack length observed on this type of alloy. These results suggest that the Vickers indenter geometry, i.e., 136°, is fairly satisfactory but not necessarily optimum for indentation cracking.

Indentation Load Levels

Surfaces of non-plastic solids, free of residual compressive stresses can be expected to form cracks at very low indentation load levels. Our laboratory research results have confirmed that Palmqvist cracks were formed in ceramics at low indentation loads (5). Similar results were observed by Lankford in ceramics and glasses (6). From Equation 1 given in the aforementioned section, the quantity $Pc^{-3/2}$ must be a load-independent quantity, for the well-behaved ceramic materials, i.e., the solid does not exhibit any phase change that can disturb the thermodynamics of the local region during indentation process. Our research results found that $Pc^{-3/2}$ is load dependent for tetragonal zirconia polycrystals (TZP-2 mole% Y_2O_3). This is due to the stress-induce tetragonal-to-monoclinic phase transformation that occurs during indentation process (4).

Local Microstructure

Local microstructure can affect the indentation cracking process and eventually results in meaningless crack length measurements especially at low indentation loads. There are many problems associated with applying very low indentation loads. For example, the relationship between crack length, c, and indentation load, P, can be disturbed due to local variations in microstructural features, e.g., second phase dispersoid like fibers and platelets. Secondly, the ratio c/a, or *l*/a, are normally lowest at low indentation loads which means that a greater proportion of the crack will be within the dimension of the deformed region of the impression. When c/a is much less than 1, the crack length c is meaningless. However, if the grain size is greater than 10x the crack length, c, therefore indentation cracking can be considered of a single crystal indentation type and the results will be meaningful.

Edge and Sample Thickness

Performing indentation near another impression or near the edge of the sample can have a detrimental effect of the crack shape and length. The presence of another impression and indentation crack or near the free edges of the sample can be expected to influence the stress field around the new indentation. In WC-Co hard metal alloys, it was found that with decreasing the distance of the impression from the edge of the sample, the length of the cracks parallel to the edge increased while the perpendicular cracks were unchanged. This means that the toughness determined from average crack length would be underestimated. As a rule of practice, a distance of 10 times the impression diagonal from the free edges of the sample was found to be sufficient to eliminate such an effect. Similar interactions between neighboring indentations have been already noticed where uneven crack lengths were measured as a result of stress fields surrounding the indentation impressions and cracks. Therefore, a "safe" distance that would eliminate the effects of such interactions must be determined as a function of impression size and or crack length (or indentation load).

Crack Shape and "Crack Length" Measurement

Indentation cracking initiates as a result of applying a load, P, that is equal to or more than the threshold. Under "identical conditions" equal cracks initiate from the corners of the indentation impression. Such identical conditions seldom exist and subsequently, non-equal cracks and different crack shapes initiating from the sides and the corners of the impression are usually obtained. Several crack shapes were observed throughout our extensive research regarding the application and the validity of such a fascinating testing technique. These include the following:

1. Side cracks with different crack length, i.e., crack branching.
2. Discontinuous cracks either from the corner of the impression or from its sides.
3. Non-symmetric cracks, e.g., one or two or three cracks, instead of four.
4. Non-equal crack length.
5. Angle cracking. i.e., not parallel to the major axes of the indentation impression.
6. "Butterfly" cracking and chipping.
7. Fuzzy and undistinguished impression and cracks, as a result of microstructure features.
8. Crack deflection and "zigzag"-like crack shape.

Usually the crack length is simply defined as the projected length on to a line parallel to the impression diagonal. This usually underestimates the true crack length but it is consistent with the macroscopic nature of the first mode of fracture, K_{IC}, i.e., phenomena such that crack defection and crack branching contribute to an increase in fracture toughness of the materials or lesser crack length (7).

Environment

It is well documented that a significant portion of indentation crack growth takes place during the unloading of the indenter, driven by residual stresses around the impression. It has been also documented that indentation cracks continue to grow in humid atmosphere after unloading is complete (8,9). This phenomenon was observed in silicon nitride and silicon carbide-based ceramic composites, due to what we have called Stress Corrosion Assisted Crack Propagation or SCACP. Carrying out indentation under inert environment using a mineral oil film has eliminated such a post indentation crack growth. Similar behavior was observed in glasses and was reported by different authors (8,9).

SOLUTIONS AND RECOMMENDATIONS

Should we decide to standardize such a simple testing technique to measure the hardness, fracture toughness, K_{IC}, slow-crack growth, fracture behavior of a specific brittle solid under certain conditions, etc., the following summarizes some of the solutions that were obtained from our extensive research activities in the area of Vickers indentation on many ceramic materials:

a. Sample surface preparation, i.e., grinding and polishing, must assure that there are no residual stresses that will interact with the indentation crack growth and the chemistry and physical properties of the finished surface indeed represent the bulk material, unless some surface modifications, i.e., ion implantation, chemical and/or mechanical modification are introduced intentionally for a specific application.
b. Controlling the environmental conditions under which the indentation process is carried out, i.e., humidity and temperature, to avoid any post-indentation crack growth attributed to environmental effects.
c. If the fracture toughness, K_{IC}, is sought to be calculated from the crack length measurement, c, and indentation load, P, it is highly recommended to develop a model that can incorporate the crack shape, i.e., median or half-penny, elliptical, Palmqvist or lateral cracks, etc., thus assuring the material is "well-behaved" under such loading conditions and, more importantly, the quantity $Pc^{-3/2}$ is load-independent.
d. In case of phase transformation that takes place due to stress application, e.g., tetragonal-to-monoclinic phase transformation in zirconia ceramics, it is required to develop a model than can accommodate the thermodynamics of such a transformation process.
e. Regarding the effects of local microstructure on crack shape and crack length measurement, it is recommended that finite element analyses are required to understand how the crack dissipates its energy during its propagation throughout a heterogeneous microstructure. e.g., fibres and platelets. Such analyses can provide an in depth understanding of

indentation fracture mechanics which can be correlated to fracture mechanics using DCB and SENB analyses.

DISCUSSIONS

The following is the summary of the discussions that were carried out at the end of every session and in the Open Forum. There was unanimous agreement between all the attendees of the symposium concerning the following important issues:

a. Surface finish must be carefully prepared to avoid any residual stresses that can underestimate the crack length measurements.
b. What should we do when chipping occurred during and after indentation? Why and how does it happen? What crack length should we use?
c. The inconsistency of the so called "the general empirical formulae" to calculate the fracture toughness of different ceramic and brittle solids is an important issue. Therefore, unless we have a "standardized" test conditions and a unified equation to calculate the toughness value of the material in question, a comparative crack length should be considered to evaluate the toughness qualitatively, i.e., in comparison with crack length measurements of a standard and well-known material.
d. Environmental effects on post indentation crack growth have to be controlled.
e. The effects of texture and microstructure of multi-phase ceramic composites (CMC's) on indentation behavior (i.e., crack length and crack shape), must be thoroughly investigated to obtain reliable information about the fracture behavior of these new classes of engineering materials.
f. The variability in crack length measurements for the same materials, same surface finish, same specimen for the same indentation load has to be thoroughly analyzed. Should we check the microstructure? Should we carry out a Weibull analysis?
g. How can we analyze crack branching, side cracks, discontinuous cracks, etc?

The above issues reflect the great interest of the ceramic community in such a fascinating technique that is simple, fast, easy to perform, can be done on small samples and above all, does not require the tedious machining of large size specimens. Unfortunately, little efforts were spent to look at these serious problems, but we expect that in the near future we will find solutions to the above-mentioned problems that will assure us the reliability of micro indentation technique in our materials research and development

ACKNOWLEDGEMENT

The author expresses his sincere thanks to the attendees of this symposium who contributed significantly to this open forum through the discussion periods that were carried out after every session.

REFERNCES

1. A G Evans and E A Charles, "Fracture Toughness Determination by Indentation," J Am Ceram Soc, Vol. 59, No. 7-8, pp. 371-2 (1976).
2. Ahamd G Solomah, "Grinding and Surface Finishing Effects on t->m Phase Transformation in Zirconia-Toughened Ceramics," Unpublished Results (1992).
3. Ahamd G Solomah, "x-Ray Diffraction Analyses of Ground and Polished Surfaces of Zirconia-Toughened Ceramics," Unpublished Results (1992).
4. Ahmad G Solomah, "Accuracy of Empirical Formulae to Measure the Toughness of Transformation-Toughened Ceramics- How accurate do you need your KIc?", Ceramic Transaction Vol. 156, pp. (2003).
5. Ahmad G Solomah, "Fractographic Analysis of Yttria-Tetragonal Zirconia Polycrystals (Y-TZP 2mole% Y2O3).", Unpublished Results (1992).
6. M T Laugier, "New Formula for Indentation Toughness in Ceramics," J Materials Science Letters 6)1987) 355-356.
7. Ahamd G Solomah and Leonardo Esposito, "Indentation Fracture of Silicon Carbide Whisker-reinforced Silicon Nitride Ceramic Matrix Composites (CMC's)," Ceram Engg & Sci Proc, July-August 1992, pp. 712-721.
8. G R Anstis, P Chantikul, B R Lawn and D B Marshall, "A Critical Evaluation of Indentation Technique for Measuring Fracture Toughness: I. Direct Crack Measurements," J Am Ceram Soc 64 (9) 533-8 (1981).
9. A G Solomah, A Tucci and L Esposito, "Slow-Crack Growth Study in Whisker-Reinforced Ceramic Matrix Composites (CMC's)," Ceram Engg & Sci Proc, Sept-Oct 1997, pp.

KEYWORD AND AUTHOR INDEX

9 781574 982121